POLICY AND PRACTICE IN

NUMBER FIFTEEN

CONVERGENCE OR DIVERGENCE?
INITIAL TEACHER EDUCATION IN SCOTLAND AND ENGLAND

POLICY AND PRACTICE IN EDUCATION

1: Lindsay Paterson, *Education and the Scottish Parliament* (out of print)
2: Gordon Kirk, *Enhancing Quality in Teacher Education* (out of print)
3: Nigel Grant, *Multicultural Education in Scotland* (out of print)
4: Lyn Tett, *Community Education, Lifelong Learning and Social Inclusion* (second edition, 2006)
5: Sheila Riddell, *Special Educational Needs* (second edition, 2006)
6: J Eric Wilkinson, *Early Childhood Education: The New Agenda* (2003)
7: Henry Maitles, *Values in Education: We're All Citizens Now* (2005)
8: Willis Pickard and John Dobie, *The Political Context of Education after Devolution* (2003)
9: Jim O'Brien, Daniel Murphy and Janet Draper, *School Leadership* (2003)
10: Margaret Nicolson and Matthew MacIver (eds), *Gaelic Medium Education* (2003)
11: Gordon Kirk, Walter Beveridge and Iain Smith, *The Chartered Teacher* (2003)
12: Jim O'Brien, *The Social Agenda of the School* (due 2007)
13: Ann Glaister and Bob Glaister (eds), *Inter-Agency Collaboration: Providing for Children* (2005)
14: Mary Simpson, *Assessment* (2005)
15: Ian Menter, Estelle Brisard and Ian Smith, *Convergence or Divergence? Initial Teacher Education in Scotland and England* (2006)
16: Janet Draper and Jim O'Brien, *Induction: Fostering Career Development at All Stages* (2006)
17: Sheila Riddell and Lyn Tett, *Gender and Teaching: Where Have All the Men Gone?* (2006)

POLICY AND PRACTICE IN EDUCATION

SERIES EDITORS
JIM O'BRIEN and CHRISTINE FORDE

CONVERGENCE OR DIVERGENCE?
INITIAL TEACHER EDUCATION IN SCOTLAND AND ENGLAND

Ian Menter
Chair of Teacher Education
Faculty of Education, University of Glasgow

Estelle Brisard
Lecturer in Education, University of Paisley

Ian Smith
Dean of the School of Education, University of Paisley

DUNEDIN ACADEMIC PRESS
EDINBURGH

Published by
Dunedin Academic Press Ltd
Hudson House
8 Albany Street
Edinburgh EH1 3QB
Scotland

ISBN 1 903765 47 1
ISBN 13: 978-1-903765-47-0
ISSN 1479-6910

© Ian Menter, Estelle Brisard and Ian Smith 2006

The right of Ian Menter, Estelle Brisard and Ian Smith to be identified as the authors of this work has been asserted by them in accordance with sections 77 and 78 of the Copyright, Designs and Patents Act 1988

All rights reserved.
No part of this publication may be reproduced or transmitted in any form or by any means or stored in any retrieval system of any nature without prior written permission, except for fair dealing under the Copyright, Designs and Patents Act 1988 or in accordance with the terms of a licence issued by the Copyright Licensing Society in respect of photocopying or reprographic reproduction. Full acknowledgment as to author, publisher and source must be given. Application for permission for any other use of copyright material should be made in writing to the publisher.

British Library Cataloguing in Publication Data
A catalogue record for this book is available from the British Library

Typeset by Makar Publishing Production
Printed and bound in Great Britain by Cromwell Press

CONTENTS

	Series Editors' Introduction	*vi*
	Acknowledgements	*vii*
	Glossary	*viii*
	Introduction	*x*
Chapter 1:	The study of initial teacher education as a 'home international'	1
Chapter 2:	Policy trajectories in Scotland and England	18
Chapter 3:	From policy to practice? The question of professional knowledge	36
Chapter 4:	From practice to policy? The problem of partnership	55
Chapter 5:	Convergence or divergence?	73
	Bibliography	84
	Index	90

SERIES EDITORS' INTRODUCTION

This volume focuses on the academic and professional preparation of teachers, a matter that has long been a contentious issue for successive governments and policy makers at UK level in an era of standards, benchmarks and the extension of routes into teaching, particularly in England but resisted so far in Scotland. The political and policy debate has been characterised by the perceived polarities of initial teacher training (ITT – England) or initial teacher education (ITE – Scotland). The differences in values and approach may be summed up in these two descriptors and reflect some of the views held in the two systems.

In this book, Ian Menter, now at the University of Glasgow and his co-authors, Estelle Brisard and Ian Smith of the University of Paisley, draw on a series of projects and studies conducted over the past few years. These are associated with consideration of the influences on the reconstruction of teaching, including issues such as supply and demand, temporary teaching posts and how best teachers may be prepared for the demands of new curricula and pedagogy. Using the comparative concept of a 'home international' study, the authors interrogate the patterns of policy development in Scotland and England to determine what variations in policy and emphasis exist and why these have occurred. The authors take a three-strand approach, looking at historical foundations, policy and practice. They look retrospectively at developments in Initial Teacher Education and Training (ITET) from the 1980s and how these might continue to inform or develop in the future. They assess, compare and contrast specific policies in each jurisdiction and consider the role and influence of major stakeholders. Their examination of practice involved field engagement with various processes associated with the education and training of new teachers and rich illustrations are provided. Important areas of concern for all those involved in initial teacher preparation are considered such as the nature of professional knowledge, partnership, and the role of higher education. This book offers important insights into policy associated with the development of teacher education and contributes to our understanding of how comparative approaches can provide insights into our own particular situations.

Dr Jim O'Brien
Vice Dean and Director,
Centre for Educational Leadership,
Moray House School of Education,
The University of Edinburgh

Dr Christine Forde
Senior Lecturer in Educational Studies
The University of Glasgow

ACKNOWLEDGEMENTS

Many people have played a part in supporting us while we undertook the studies that have given rise to this book, as well as during the writing process itself.

The study was supported and initially funded by the University of Paisley Research Committee. Excellent secretarial support was provided by Sheila Boyd and Jean Paton at the University of Paisley and by Susan Gully at the University of Glasgow.

We were fortunate in establishing a superb advisory group comprised of two Scots and two English people: Professor Gordon Kirk, Ms Myra Pearson, Professor Pat Mahony and Ms Mary Russell. Our thanks go to them for their guidance and support, although they should not be held to bear any responsibility for what follows in this volume.

We also offer a general thank you to all the unnamed institutions and members of the teacher education communities in England and Scotland who supported our work by taking part in this study, all of them busy people who gave willingly of their time.

In the book we have drawn from a number of articles and conference papers that have been written and presented by us over the past five years, including work done for the General Teaching Council for Scotland on models of partnership in initial teacher education. These are referred to in the main text of the book at appropriate points.

Ian Menter, Estelle Brisard and Ian Smith
May 2006

GLOSSARY

BA/BSc	(with TQ): Bachelor of Arts/Bachelor of Science, with Teaching Qualification
BEd	Bachelor of Education
CATE	Council for the Accreditation of Teacher Education
CNAA	Council for National Academic Awards
CPD	Continuing Professional Development
DES	Department of Education and Science
DfE	Department for Education
DfEE	Department for Education and Employment
DfES	Department for Education and Skills
EIS	Educational Institute of Scotland
EBR	Employment-Based Route
GTCE	General Teaching Council for England
GTCS	General Teaching Council for Scotland
GTP	Graduate Teacher Programme
HE	Higher Education
HEI	Higher Education Institution
HLTA	Higher Level Teaching Assistants
HMI	(or HMIE for Scotland): Her Majesty's Inspectorate (of Education)
ICT	Information and Communications Technology
ITE	Initial Teacher Education (Scotland)
ITT	Initial Teacher Training (England)
ITET	Initial Teacher Education and Training
LEA	Local Education Authority
NQT	Newly Qualified Teacher
OECD	Organization for Economic Cooperation and Development
OfSTED	Office for Standards in Education
PGCE	Postgraduate Certificate in Education
PGDE	Professional Graduate Diploma in Education (Scotland)
QAA	Quality Assurance Agency for Higher Education
QCA	Qualifications and Curriculum Authority
QTS	Qualified Teacher Status
SCITT	School-Centred Initial Teacher Training
SCOP	The Standing Conference of Principals
SEED	The Scottish Executive Education Department
SEN	Special Educational Needs

SFC	Scottish Funding Council (for Further and Higher Education)
SHEFC	The Scottish Higher Education Funding Council (replaced in 2005 by SFC)
SITE	Standard for Initial Teacher Education (Scotland)
SOED	Scottish Office Education Department
SOEID	Scottish Office Education and Industry Department
STEAC	Scottish Tertiary Education Advisory Council
STEC	Scottish Teacher Education Committee
TDA	Training and Development Agency for Schools
TTA	Teacher Training Agency
TQA	Teaching Quality Assessments
UCET	Universities Council for the Education of Teachers

INTRODUCTION

The purpose of the book is to explore the similarities and differences in the trajectories of policy and practice in initial teacher education and training (ITET) in Scotland and England, especially in the post-devolution context.

The book draws mainly upon a two-year study undertaken at the University of Paisley by the three authors. This study has generated an extensive set of data, largely consisting of interview transcriptions. During the study a total of 27 policy makers were interviewed on either side of the border and a total of eight sites of practice (three in Scotland and five in England) were examined. Observations of teaching and learning were carried out and 90 interviews were conducted with professional staff involved in the provision of ITET (including higher education-based and school-based staff) and with students/trainees who were on the training programmes. The study also made use of a range of national policy documents as well as papers gathered from providers.

The main themes explored in the study include:

- defining the key characteristics of ITET provision, including system of governance, institutional arrangements, quality assurance processes, curriculum and assessment, as well as the significance of ITET within national systems of education, including a consideration of the extent to which ITET influences the wider system – or vice versa;

- the professional contexts for ITET in Scotland and England, including its relationship to other aspects of teacher professional development, and the roles of key stakeholders;

- the significance of the recent historical, political, social and cultural national contexts for the development of ITET policy and practice;

- the extent to which developments in ITET in these two countries demonstrate increased convergence or divergence (perhaps under the influence of globalisation or of post-devolution nationalism).

The last of these points gives rise to the title of this book and focuses our attention on what we have found, in the course of this study, to be a deeply fascinating but complex question. Scottish education policy has been distinctive throughout the history of the United Kingdom. Throughout most of the twentieth century, education policy developed in Westminster by the UK government has been variously adopted, adapted, subverted or rejected by Secretaries of State for Scotland, their junior ministers and the civil servants of the Scottish Office. These decisions, made in Edinburgh, have often been influenced by a range of key stakeholders in Scottish education, including teachers, their unions, the General Teaching Council for Scotland (GTCS), local authorities and the inspectorate. Accounts of this fascinating history can be found in the work of Paterson (2003), Anderson (1995) or Holmes (2000).

However, in this study, while we consider the latter part of the twentieth century in some detail, our central interest is to explore the significance (or otherwise) of the devolution that occurred right at the end of the century. These new institutions did not entirely replace their forerunners, partly because some areas of policy were 'reserved' for determination by the UK government and so, for example, a much-reduced version of the Scottish Office still exists in London and there is still a Secretary of State for Scotland – as well as a First Minister in the Scottish Parliament. Nevertheless, education policy is not a reserved matter and so, for the first time since 1707, Scotland became entirely free to determine how the provision of education should be developed. It was this new independence which gave rise to our interest in the question: will this new freedom lead to (even) greater distinctiveness in Scottish education, a divergence between the two systems, or, on the other hand, will there be a move towards greater similarity, perhaps under the influence of wider European or indeed global forces, leading to convergence of policy and practices in education?

Aspects of these themes have previously been considered by a range of other authors, sometimes on a one country basis (e.g. Kirk, 2000; Paterson, 2000; Bryce and Humes, 2003; Pickard and Dobie, 2003) sometimes on a comparative, specifically 'home international' basis (e.g. Raffe *et al.*, 1999; Menter *et al.*, 2004). What makes this study and this book distinctive, and we hope original, above all, is that none of these predecessors have developed a sustained analysis of the particular case of initial teacher education and training through a comparative approach within the UK.

Our account of the study is organised into five chapters. Chapter 1 explains the background to the study and its significance in greater detail. We then set out the methodology adopted in the study and point out some of the complexities of comparative research in education, even – or perhaps especially – when the cases examined are so close to each other. Chapter 2 offers a recent historical overview of ITET in each of the two countries and examines some aspects of the relationship between policy and practice in education, setting out the view we hold of policy as a process, rather than as a product or outcome. Chapters 3 and 4 then focus respectively on two aspects of ITET which emerged from the study as areas of special interest and significance. The first area is that of 'professional knowledge' where we consider how a range of stakeholders seek to influence the ways in which teaching is defined and hence the nature of the ITET curriculum. This provides a case in which we can track the move from policy through into practice. The second area, examined in Chapter 4, is 'partnership', the term that is used to describe the relationships between the various parties involved in the provision of ITET. Here we see how the practice of ITET can have a dramatic effect on policy. Having examined these two cases, we have not only demonstrated the significance of each area, but have explored the complex relationship between policy and practice – it is in neither case actually a linear process of the one influencing the other. The final chapter offers a review and assessment of the extent to which the case of ITET demonstrates convergence or divergence in educational policy and practice in Scotland and England and returns to wider questions about globalisation, standardisation, nationalism and the role of the state and of professionals in education.

Chapter 1

THE STUDY OF INITIAL TEACHER EDUCATION AS A 'HOME INTERNATIONAL'

Introduction
This chapter explains the background to the study that is reported in this book. Firstly we explore why the study of initial teacher education and training (ITET)[1] has greater significance than might be immediately apparent. We then consider current understandings of the influence of globalisation on educational policy and practice in order to review the relevance of comparative studies at the present time. This leads into an examination of the particular dual case of Scotland and England and the significance of that case in the wake of political devolution at the end of the twentieth century. The final part of the chapter describes the methodology that was developed for the study at hand, including an outline of the range of techniques and research instruments that were developed and a summary of some of the methodological challenges that arose.

The significance of ITET as a field of study in educational policy and practice
The manner in which the next generation of teachers is prepared for entry into the profession can logically be seen as a manifestation of the priorities that have been set for education in the future. The values that new teachers are expected to espouse and to demonstrate in their work, the curriculum in which they are expected to have expertise and the relationships that they are expected to develop with pupils might all be seen as indicators of how education is expected to develop in the years ahead. This is but one reason why the study of ITET has a significance beyond the immediate. Through a reading of ITET policy and practice, we may reasonably be hoping to detect insights into the contemporary framing of any nation's wider education system.

In reality however, teacher education is not always to be found at the leading edge of developments in education policy and practice, as perhaps was well demonstrated when, during the 1990s in England, the implications of the enormous curriculum and assessment reforms being introduced into schools following the 1988 Education Reform Act were only subsequently to be brought into programmes of initial teacher training (ITT).

The shaping of ITET has also been strongly influenced in some contexts by deeply held ideological positions about the nature of teaching. Particular governments have, at one time or another, determined to remove the influence of certain vested interests and to replace that influence with their own. On some occasions this has been about reducing the influence of teachers or teacher educators (mostly about setting a national standard for provision). Furthermore, the reality is that there have been times and places in many European and North American settings where ITET policy has been affected much more by the need for speedy training and entry to the profession of particular groups of teachers – in other words, by teacher supply – than by longer term thinking and planning.

ITET policy therefore is closely bound up with political, economic and social developments in a society. This makes it a rich field for historical and sociological study, but also one that is profoundly complex. The study reported here sought to address that complexity through taking an Anglo-Scottish comparison as a case that we felt would be illuminative. This meant approaching the project from a consideration of wider contemporary aspects of educational policy and practice.

Initial teacher education and training in Scotland and England in the age of globalisation

Our project, 'Convergence or Divergence? Initial teacher education in Scotland and England', originally stemmed from an assumption that there may exist 'unique cultural configurations' (Epstein, 1992, p. 13) in the ways in which ITET policy and practice are conceived, organised and conducted in these two countries. Using a comparative approach, the intention was to test for 'national uniqueness' in the way ITET policy and practice take place in England and Scotland, as well as to look for prevalent patterns, which transcend national specificity (Smyth *et al.*, 2001, p. 28). Another assumption was that the complex set of influences on UK countries generated by political and macro-economic phenomena such as devolution, the process of European integration and globalisation is likely to result in diverse patterns of policy convergence and divergence in education and teacher education (Paterson, 2000; Pickard and Dobie, 2003; Bryce and Humes, 2003).

While some note a divergent trend in educational policy among the four UK education systems (Raffe *et al.*, 1999), there has been a growing literature in education policy studies about the convergence of education policies and practices which is frequently associated with the thesis of globalisation (Adick, 2002; Arnove and Torres, 1999; Green, 1999; 2002). Globalisation refers to 'the *processes* through which sovereign national states are crisscrossed and undermined by transnational actors with varying prospects of power, orientations, identities and networks' (Beck, 2000, p. 11, quoted in Jarvis, 2002, p. 7). This new relationship between nation states and the world economy affects nation states' ability to make national policy independently (Dale, 1999, p. 2). For Ozga (2005), globalisation redefines and reinforces the links between education and the economy thereby shaping education policy and driving the modernisation agenda in education. Cross-national convergence in educational policy has been stimulated by a new educational consensus in the form of macro-policy solutions disseminated by supranational organisations like the Organization for Economic Development (OECD) and increasingly since the Maastricht treaty of 1992, the European Union (Ball, 1998; Henry *et al.*, 2001; Lawn and Lingard, 2002; Nóvoa, 2000). In teacher education specifically, some also account for a trend of international convergence since the 1990s by the fact that around the world, technocratic rationality and technicism have been denounced in ITET programmes, forcing teacher educators 'to find a new way to integrate harmoniously the critical, the academic and the professional components of a teacher education programme' designed to prepare teachers to become critical intellectuals (Schnaitmann, 1998, p. 156).

In the case of ITET in Scotland and England, there are features of recent developments which on the face of it seem to be very similar. One is the development of competence standards for entry into the profession. This development is a response to a perceived need, in the 1980s and 1990s, both in North America and Europe, to control the quality of performance of teachers and educational systems and can be linked to wider trends towards increased *accountability* in public services (Ryba, 1992; Merryfield, 1994; Mahony and Hextall, 2001). Another similar development is the formal acknowledgement of the significant role of schools in the ITET process, usually signified through the use of the term *partnership* (see Brisard *et al.*, 2005). European countries have tended to deal with common concerns with quality improvement and recruitment in initial teacher education (ITE) by applying similar solutions, thereby giving an impression of increasing convergence. According to a recent Eurydice report, 'the desire to improve the quality of training is usually reflected in its acquiring a more professional focus' (*professionalisation*) as well as in 'making provision more uniform' (*rationalisation*) (Eurydice, 2002, pp. 17–19). Professionalisation generally refers to a process of enhancement of the social and professional status of teaching and teachers by making the ITE provision more professionally oriented (i.e. 'theoretical and practical teacher training occupying a more prominent position', p. 19) whilst at the same time upgrading its academic status (i.e. ensuring 'more demanding requirements as regards the content of learning', p. 19). The professionalisation agenda was often associated with measures to make ITE requirements – with regards to the content and organisation of courses – more uniform. Rationalisation meant, for instance, blurring out differences in status and preparation between teachers inherited from the historical evolution of national education systems in Europe. In practice, it meant consolidating all programmes of ITE within the higher education sector and preferably in one single structure or type of institution. One widespread option was the progressive incorporation of teacher education in the mainstream of university activity (*universitisation*), which took place over different phases and durations in different countries, while some countries retained or introduced a higher education provision outside the university (Galton and Moon, 1994). The international trends in the development of initial teacher education and training over the past 40 years identified above (*accountability*, *rationalisation*, *universitisation*, *professionalisation* and *partnership*) constitute five key themes in the comparative study reported here. We will come back to these themes in reference to the specific contexts of Scotland and England in Chapters 2, 3 and 4, as well as in our concluding analysis of the current pattern of convergence or divergence in ITET policy and practice between the two countries in Chapter 5.

But the complexity of the system of influence currently at play is perhaps best demonstrated by the fact that while a trend towards convergence has been promoted by the European Commission and OECD policies, systems of teacher education in Europe nevertheless remain of 'a highly heterogeneous nature' (Buchberger *et al.*, 2000, p. 12). Although globalisation does 'alter the prospects for traditional education systems' with new global policy rhetorics emerging (e.g. lifelong learning), education systems are not converging on a single model (Green, 2002, p. 14). This is because common prescriptions by international agencies in the field of education and global forces 'are not uniformly implemented or unquestionably received' (Arnove, 1999, p. 2). With regards more specifically to ITET in Europe, Gauthier (2002) reported that despite a convergence in the nature of the challenges in this domain, solutions nevertheless remain specific and in keeping with the cultural tradition of education of the host countries.

Far from making cross national research irrelevant or meaningless, it is our belief that the complex system of influences generated on UK countries by macro phenomena such as devolution, the consolidation and expansion of the European Union and globalisation provides a particularly rich setting for research into the relationship between the state, the nation and education. Recently published findings report important variations between England and Scotland's approaches to the process of modernisation of the profession (Menter *et al.*, 2004). Similarly, Ozga (2005), drawing on the findings from the Education, Governance and Social Integration and Exclusion in Education (EGSIE) project (see Popkewitz *et al.*, 1999) argues that political devolution may support increased policy divergence between England and Scotland because of differences in their 'collective narratives' (defined as the way policy and practices reflect the cultural identity of a country). Crossley and Watson also note that 'a rationale for the resurgence of local cultural movements can, paradoxically, be found in the globalisation process and in the weakening of the Nation State' (2003, p. 57). In Scotland, the education system, along with the church and the legal system, is seen as one of the famous 'three pillars' of Scottish identity and as a key part of the 'nation building' which is going on in a double context of globalisation and devolution. In this context, teachers may be seen as 'nation builders' and the ways in which they are trained may be expected to reveal some distinctive features which relate to this process. It is against this background that the 'Convergence or Divergence study' sought to develop an understanding of the connections between initial teacher education and aspects of nationhood.

Rationale for a 'home international' study of ITET

The establishment of the Scottish Parliament and the Scottish Executive in 1999 signified the greatest constitutional change in the relationship between England and Scotland since the creation of the British State by the Treaty of Union in 1707. Prior to the devolution of powers from Westminster to the Scottish Parliament, the UK government would develop policy for England, and would expect to see these developments mirrored in Wales, Northern Ireland and Scotland (Paterson, 2003). The Secretary of State for Scotland was responsible for developing policy, although in practice education was usually the day-to-day responsibility of one of the ministers within the Scottish Office. The pattern of policy development and implementation was a mixture of leadership (where England was the main leader), autonomous policy making (to pursue domestic agendas) and reactive policy making (whether to delay, accept, adapt or reject initiatives originating in Westminster) (Raffe, 1998). Following devolution, the UK Parliament at Westminster has devolved powers to the Scottish Parliament in Edinburgh and the Scottish Executive is now directly responsible for a number of devolved matters such as education, health and the law among others.

Raffe *et al.* (1999) have eloquently argued for the value and feasibility of undertaking comparative research into what they call 'home internationals': England, Scotland, Northern Ireland and Wales. The comparative potential arises from their historical constitutional interdependence which brings about a meaningful balance of similarities and differences in the way education policy was made and implemented in these countries, at least until devolution. There are differences between the territories however. Right up to 1997 it was generally the case that Westminster policies were followed through very closely in Wales and Northern Ireland, although often with specialised additions such as the establishment of Welsh in the curriculum, Welsh medium teaching in some schools, or the introduction of a curriculum theme 'education for mutual understanding' in Northern Ireland and other variations on the 'follow-my-leader' principle (Raffe et al., 1999, p. 10). In Scotland however, there was often considerable resistance to the adoption of Westminster derived policies resulting in either major adaptations or, occasionally, complete rejection. Two recent examples of this would be the very different approach taken to curriculum and assessment in the two countries (see Brisard and Menter, 2004) and the varying take-up of formal self-government of schools.[2]

A principal advantage of UK-wide studies of education systems, and one which is noted by Raffe (1998), is first of all to promote a better knowledge and informed understanding of the similarities and differences within the systems, which are still too often wrongly amalgamated into 'the English education system'. Another important benefit is that they illuminate current common issues and tensions within specific areas of education and training, and highlight a number of alternative ways in which the four components approach these issues – what Raffe et al. (1999, p.18) call 'variations upon common themes' – and importantly, the influencing factors beneath these variations.

To summarise, there appear to be three ways in which developments in England and Scotland may be portrayed and understood. The first might be described as *parallel development*. Here progress is seen to occur through policy borrowing and /or sharing between nations. In this view, we become particularly aware of the similarities between the two countries under consideration. Examples would include the processes of teacher certification and programme accreditation or the moves towards basing ITET in universities. The second approach, the *UK specific approach* recognises the tensions and conflicts – historical and contemporary – that have influenced each part of the UK in different ways. Even in a historical context of constitutional interdependence and recently, a wider process of Europeanisation and globalisation, this view emphasises differences, for example the retention of a commitment to university involvement in ITE in Scotland in contrast to the introduction of non-Higher Education (HE) routes of entry in England. The third approach, under the heading of *globalisation*, emphasises the influence of transnational organisations such as the OECD, for example, in introducing market forces, or standards and accountability, albeit in different forms, into all education systems, including teacher education, across the 'developed' world. Each of these three perspectives has influenced the shaping of this study. Indeed, to some extent, the study may actually be seen as an attempt to assess the relative significance of each one.

The Convergence or Divergence study: conception and underpinning principles

The project reported here sought to identify the 'pattern of convergence and divergence' (Keating, 2002, p. 3) in the policy making process in ITET as well as to map out the nature and organisation of the training provision in Scotland and England, at a time of important constitutional change in the UK. Our aim has been to develop an approach that can provide insights into the significance of similarities and differences between two geographically and politically contiguous education systems. Although some argue that the differences between the teacher education systems of nation states are superficial and mask underlying and much more significant similarities (Sander, 2000), what emerges in the forthcoming chapters is that the differences identified between England and Scotland are key signifiers of cultural, social and political variation, notwithstanding a large number of similarities.

Following a number of principles established by the work of leading comparative educationists, we adopted a socio-historical approach to comparative research which we hoped would allow for the research object(s) to be considered in a dynamic perspective (Popkewitz *et al.*, 2001, Alexander, 2000; Crossley and Watson, 2003). A central principle of our approach, and one which is advocated by Crossley and Watson (2003, p. 142) is 'context sensitivity'. Citing Sadler and Lauwerys respectively, they note that 'every system of education is shaped by its local, historical, economic, cultural and social context' and that its underlying philosophy cannot be truly understood without this fundamental principle being acknowledged (p. 39).

A second underpinning principle of our comparative approach is that it 'should be powerfully informed by history' (Alexander, 2001, p. 5). In other words, if it is to identify what direction ITE policy making is likely to take under devolution in Scotland, this study had to address, in its design, the need for a careful understanding of what the past pattern of convergence and divergence in ITE policy making is between England and Scotland (Keating, 2002). Raffe *et al.* (1999) also stress the importance of having 'a historical perspective and a capacity for analysing social change' if we are to understand what they see as a current divergent trend among the four UK education systems. Taking into account the dynamic nature of the processes under study however, meant going beyond an analysis of systems and structures in order to give some consideration to the experiences and practices of key actors working at the policy level, or at the institutional level. Although national and institutional policy frameworks for education need to be taken into account, the comparison cannot afford to neglect the role and capacity for action of the actors involved, who can influence or even subvert policy in the process of implementation (Osborn and McNess 2005; Moreau, 2005). We believe it is important to illuminate participants' own perspectives as a key variable in understanding teacher education practices in different national settings. It is how they see and interpret the situations which interests us rather than 'how that world appears to the outside observer' (Blumer, 1971, p. 21). As such, a third guiding principle of our study was a particular focus on the discourses (text-based or actor-based) employed across policy contexts or across contexts of practice and the connections between these discourses and the institutional structures in which they take place. In other words, during the data collection process the emphasis was placed on the lived experiences (accounts) of key people 'inside the system', whether they be policy makers or teacher educators, in addition to the analysis of administrative, policy or institutional documentation.

We agree that attention needs to be paid, through comparative educational research, to the impact of 'differing cultural perspectives on policy priorities' and we acknowledge the significance of qualitative and context-sensitive forms of comparative educational research that 'engage with a diversity of stakeholders and cultures at grass roots level' (Crossley and Watson, 2003, p. 138). In order to best take into account these various principles, we organised our research project around three strands: historical, policy and practice. Our intention was to examine the interplay between the three, before we would claim to be able to carry out a deep analysis. While we do not claim that our data gathering techniques are unique in themselves, we have sought to ensure that the combination of approaches taken has enabled us to develop both intra- and inter-nation analyses which are original and valuable.

A three-strand approach: historical, policy and practice

The historical strand examined the parallel development of ITET (policy and practice) in each country. Our study is end-loaded towards the development in ITET which happened since the 1980s in both countries. The policy strand on the other hand focused on the period from the late 1990s onwards, looking at the organisation and governance of ITE in each country. The institutional stakeholders whose roles were examined include:

> In Scotland: The Scottish Executive and its Education Department, the Inspectorate (HMIE), the General Teaching Council for Scotland (GTCS), Local Education Authorities (LEA) and ITE providers.

> In England: The Department for Education and Skills (DfES), the Teacher Training Agency (TTA), the General Teaching Council for England (GTCE), the Office for Standards in Education (OfSTED) and ITT providers.

Both the historical and the policy strands are very closely intertwined as data generated in the context of the first strand was used to inform the second. Data about the past and present pattern of policy making in ITET in England and Scotland and about the resulting developments in content and structure were collected from official documentary sources as well as from semi-structured interviews with former and current holders of significant positions in the institutions mentioned above. Interviews with retired members of key stakeholder institutions were particularly useful to examine the history of policy making and 'its accompanying narrative of explanation of education policy' (Ozga, 2000, p. 114). McPherson and Raab (1988) and Humes (1986) have respectively advocated interviews as the main method of data collection in educational policy research, arguing that other types of data (primary or secondary literature for instance) can provide information about the 'what', but not the 'why and how' of policy making as a process and that interviewees can provide 'insider' anecdotal information of the way initial teacher education policy making and practice take [took] place in their country/institution at a given time.

Beyond a focus on systems and policies, we felt that studies which take practice as 'the most appropriate unit of analysis in understanding education, learning and developments cross-culturally' can positively influence comparative education research (Hoffman, 1999, p. 483).

The practice strand aims to offer a rich description of current practices in ITET in both national contexts. In taking a context-sensitive approach, we wished to consider how practice 'relates to the context of culture, structure and policy in which it is embedded' (Alexander, 2000, p. 3). For this purpose, a multi-site, ethnographic approach to data collection and analysis was adopted (see below), which involved looking at eight ITET providers in both contexts: five in England, and three in Scotland.

A long-standing difficulty in comparative research is the question of the selection of suitable units of comparison. In a study such as this one, a first level of comparison is intra-national (across the sites visited within one country) and a second level is cross-national. Walford (2001) stresses the need to give sufficient concern to – and to offer some theoretical justification for – the selection of research sites, over and beyond the primary consideration of obtaining access. A key starting point when designing the practice strand of the study was the fact that all teachers who trained in Scotland are qualified to work south of the border but the reverse is not true of their English-trained counterparts. Indeed, in the autumn of 2001, the General Teaching Council for Scotland turned down the applications for full registration from two teachers who had trained through the Graduate Teacher Programme (GTP) in England on the ground that:

> when people apply for registration, we look at whether they have undertaken a professional education programme *comparable* with the pattern of initial teacher training provision in Scotland. The three key areas are pedagogy and professional studies; subject studies; and experience in schools. As far as the GTP courses those two candidates had undertaken are concerned, we felt our requirements had not been met in the first two: professional studies and pedagogy, and subject standards. (Thornton and Munro, 2001, p. 4)

What this reflects, it seems, is diverging conceptions of what teaching – and consequently initial teacher education – is or should be about. Such contextual knowledge prompted us to pay particular attention, when selecting research sites for the practice strand, to the issue of intra-national variety in the type and models of provision on offer.

As such in England we selected an 'old university' provision (postgraduate programmes for primary and secondary), a 'new university'[3] provision (postgraduate and undergraduate routes, as well as some flexible routes), a Graduate Teacher Programme (employment-based route), a School-Centred Initial Teacher Training provision (secondary ITT only) and a college of education provision (postgraduate and undergraduate routes). In Scotland, the research sites consisted of three of the seven available university departments of education which offer ITE courses.

Data collection

The research team identified a list of 'themes' (Table 1) which were presented in advance of interview to both the historical and current policy interviewees to facilitate and structure the interview process.

The themes were arrived at partly as a result of reflection by the authors on general academic writing on the nature of teacher education, and more specific writing on the historical context for ITET in Scotland and England. For example, in relation to Theme 1, Tomlinson (2001, p. 107) highlights the particular problems in recent and current recruitment to the teaching profession in England. Specific initiatives such as financial incentives for new entrants are mentioned, as well as more 'flexible' forms of 'school-based' ITT compared to the models offered in Scotland. On the other hand, in Scotland, Marker (2000, p. 291) makes looser references to recruitment issues reflecting the fact that overall recruitment levels in Scotland are healthier than in England, at least sufficiently so that central government has not felt the need to launch specific initiatives such as financial inducements for new entrants. Using Theme 1 (with its expanded sub-themes), we were interested to see whether any Scottish interviewees would suggest that they viewed recruitment to ITET as healthier in Scotland than England because Scottish schoolteachers have been more highly valued within the national culture, or because the Scottish economy has provided fewer alternative opportunities for certain types of graduates wishing to work locally.

In addition to the themes, interviewees within the historical strand were provided with two sets of 'Key Moments in ITET' since the 1960s (Table 2), one for Scotland and one for England, regardless of which system their own career background was in. Interviewees were not expected to make explicit comparisons however, and the team subsequently analysed the transcript of interviews for evidence around convergence and divergence between England and Scotland. Scottish historical interviewees were more likely to

Table 1: 'Themes' and sub-themes for historical and current policy interviewing

Themes	Associated sub-themes
1. Relationship between the formation of teachers through ITET, national identity and the cultural and economic contexts	- National cultural traditions of education including attitudes to education - Labour markets in Scotland and England
2. Overview of the historical development of ITE in England and Scotland	- Evolution - Key aspects - Turning points
3. Scottish/English conceptions of teaching as a profession	- The professionalisation of teaching - Nature and evolution of the teacher's role, professionalism and professional knowledge - Partnership between the main parties involved
4. Nature of the teacher education process	- Types of training institution - Models of ITET - Balance between Higher Education (HE) and school-based contributions to the courses
5. Governance of ITET	- State involvement in ITET, identity and connections between the stakeholders - Balance between Higher Education Institutions (HEI), schools and local authority - Balance between open democratic processes and professional influence, ownership and control. - ITET Regulatory Framework

make comparisons between Scotland and England (even if limited), whereas English historical interviewees tended to indicate insufficient familiarity with the Scottish system to enable them to make such comparisons.

The key moments provided us with a 'definition' of the historical strand and with points of reference for the historical interviews. Again, the identification of these key moments was based upon reflection on secondary sources, but importantly supplemented by reflection on specific primary sources, especially published policy documents. Different levels of topic were mixed, with some alluding to more general developments over a period of several years within the historical framework, whereas others referred to single, specific policy documents (Table 2). However, great care was paid to achieving some consistency between the key moments for Scotland and

Table 2: Key Moments in ITET

SCOTLAND

- Teaching Council (Scotland) Act 1965: Creation of the General Teaching Council for Scotland
- Contraction in student intake and instability of forward planning in the monotechnic sector
- Debate over the future of the college of education system (mid to late 1980s)
- Significant recovery of student intakes (early 1990s)
- Specific mergers of colleges of education with Scottish universities (1990s)
- Her Majesty's Inspectors (HMI) Institutional Review (early 1990s)
- Placing of the 1993 Guidelines for Courses of Initial Teacher Education in the context of the Guidelines which existed before then
- The Teachers (Education, Training and Recommendation for Registration) (Scotland) Regulations (1993)
- The Scottish Higher Education Funding Council (SHEFC) Teaching Quality Assessments (TQA) (1994–5)
- Contraction of student intake (mid 1990s)
- Mentor Teacher Initiative (1992–5)
- Report of the GTC Working Group on Partnership 1997
- HMIE Aspect Review (2000–1)
- *A Teaching Profession for the 21st Century* (McCrone report and agreement on recommendations) (2000–3)
- Pilot Collaborative Review of Initial Teacher Education in Scotland (2001–2)
- First and Second Stage Reviews of ITE (2001–3)
- Introduction of the Teacher Induction Scheme (August 2002)
- The new Standard for Initial Teacher Education in Scotland (SITE) (2000)

ENGLAND

- Creation and development of the BEd (late 1960s)
- Closures and mergers of colleges of education (early 1970s)
- Involvement of the Council for National Academic Awards (CNAA) with the validation of courses of teacher education (from 1972–3)
- Publication of government criteria for approval of courses (1984 & 1989)
- Introduction of two new routes into teaching: the Licensed Teacher Scheme and the Articled Teacher Scheme (1989)
- Demise of the CNAA (1992)
- Kenneth Clarke's announcement of a radical shift in ITT (1992)
- Unification of the university system (1992)
- Department for Education (DfE) Circulars 9/92 & 14/93: emphasis on practical training; new partnership between HEIs and schools; introduction of School-Centred Initial Teacher Education Programmes (SCITTs) and of a competence-based model of ITE
- Creation of the Office for Standards in Education (OfSTED) (1992)
- Creation of the Teacher Training Agency (TTA) (1994)
- OfSTED framework for the inspection of ITT courses (1996, 1998)
- Government Green Paper: *Teachers: Meeting the Challenge of Change* (1998)
- Publication of Standards for the Award of Qualified Teacher Status (QTS) / introduction of curricula for ITT (1998)
- Establishment of a General Teaching Council for England (Teaching and Higher Education Act 1998)
- Introduction of Skills tests (numeracy, literacy, information and communications technology (ICT)) for all beginning teachers (from 2000)
- Publication of the revised standards for QTS (2002)

England in terms of this mix, in terms of the number of key moments identified, and in terms of the broad time frame established for both Scotland and England. The key moments were not worked through systematically during interview, rather, they were provided as an aid to interviewees in structuring their thoughts in advance of interview, and as a point of reference allowing the researchers to explore particular aspects which may not have been discussed spontaneously.

Finally, the team also identified 'Recent policy developments in ITE' for both Scotland and England which were sent to the current policy interviewees before interview and consisted essentially in the most recent items from the 'Key Moments' documents provided for historical interviewees.[4]

The practice strand of the study aimed to provide a rich description of ITET practices in England and Scotland, that is a detailed account of the various training processes observed during the researchers' immersion in the various national and institutional training contexts and an interpretation of the deeper structures which underpin the way the process of learning to teach is conceived and implemented. It is somewhat similar to Geertz's definition of ethnography as 'thick description', a notion originally borrowed from Gilbert Ryle. Geertz (1973, pp. 7–9) describes the object of ethnography as 'a stratified hierarchy of meaningful structures' and the process of analysis as 'sorting out the structures of significance'. The product of the research process as such is both descriptive and interpretative. During fieldwork, the aim was to 'get a feel' for what is going on in terms of teaching and learning practices, and for the wider organisation and management of the ITE process. In this aim, one of the team's researchers spent five days in each of the eight ITET provisions selected, to undertake both interviews with specifically identified members of the training staff and students and observations of a number of selected events in schools and within the institution (Tables 4 and 5 below). The 'ideal-typical' categories of individuals selected for interviews and programme activities identified for observation are listed in Table 3. In all cases, a balanced sample of respondents and activities was secured through advanced negotiation with a key contact person in each site.

Staff and students' individual and collective discourses and practices were seen as meaningful doorways into the ITE culture of the institution. One way the researcher *in situ* attempted to obtain a rich description of practices was through the triangulation of staff and trainee-teachers' discourses: moving between normative conceptions of the training process on the one hand and lived experiences of learning to teach on the other. Another way was through the confrontation of the researcher's observation notes with, for instance, course documentation, or staff's declared objectives and rationale for their action, which were elicited through pre- or post-observation discussion or through formal interviewing. The researcher

Table 3: Fieldwork in the practical strand

Interviews with senior staff one to one (4 or 5 interviews maximum)	Attendance at meetings	Small group interviews	Institution-based observation and follow up discussions with tutors	School visits (Ideally: 1 secondary and 1 primary setting) *Possible activities, when available, not an exhaustive list*
• Dean or Head of School of Education/ Programme manager	• Faculty or School board	• Curricular area/ subject studies lecturers	• Subject studies sessions	• Collective or individual interviews with members of school staff involved in ITET (Senior management including, where appropriate, regent, supervising teachers, mentors)
• Co-ordinator of primary ITET (whether PGCE or BEd)	• Faculty or School Executive meetings	• Educational and Professional studies lecturers	• Educational and Professional studies sessions	• Collective interview with student teachers placed in the school (in addition or instead of collective interview with student teachers in the institution)
• Co-ordinator of secondary ITET	• Quality Assurance Committee meetings	• PGCE primary/ BEd BA primary lecturers	*(to include variety between lectures, seminars, tutorials)*	• Attendance at a weekly 'subject mentor' session or equivalent
• Any other senior staff identified according to context	• Programme planning and management meetings	• PGCE secondary lecturers	• Programme delivery using ICT	• Attendance at a weekly regent or professional mentor session, or equivalent
	• Departmental/ section meetings	• Student teachers (or could be done in school)	• other sessions specific to a given provision	• Observation of student teaching and of follow-up discussion by the supervising teacher
	• Staff/student representative meetings (difficult timing)	• Link tutors (where appropriate)		• Observation of formal visit by the institution-based tutor (lesson + post-lesson discussion with HE lecturer, student and supervising teacher)
	• Partnership liaison meetings	• other ITET tutors specific to provision		

Table 4: Interview data for current practice strand

	SCOTLAND	ENGLAND	TOTAL
Institution-based	24	23	47
School-based	18	25	43
Total	42	48	90

Table 5: Observation data for current practice strand

	SCOTLAND	ENGLAND	TOTAL
Meetings	4	5	9
Seminars	9	3	12
Lectures	2	1	3
Other	1	4	5
School-based	0	5	5
Total	16	18	34

Table 6: Themes addressed during fieldwork in the practice strand

- impact of national governance and regulatory frameworks of initial teacher education and training;
- teacher education institutions' position in the Higher Education sector;
- organisational structure of the institution in delivering ITET programmes;
- underpinning conceptions of teacher professionalism and professional knowledge, model of ITET and process of learning to teach;
- teaching qualifications, routes towards qualifications, recruitment pathways;
- partnership, balance between HE and school-based contributions to the programmes;
- teaching and learning experiences for the students; student assessment;
- staffing, career structure and resourcing.

used a number of different interview schedules and observation grids, all designed around common key questions or foci for observation. The kind of themes addressed during the various interviews and observation sessions are outlined in Table 6.

We did not look to address all of these with all the interviewees, but it was left to the researcher's judgment to select, on the day, those which were considered to be particularly relevant to the individual – or group of individuals – interviewed.

Issues in the research methods employed

Semi-structured interviewing is the essential investigative approach in this project for the historical and current policy strands. We have not been attempting to achieve a full-scale life history approach in our interviews, even with our historical interviewees. Rather, like Ozga (2000, p. 126), we have been aspiring more to Bertaux's (1981) notion of 'representativeness'. Recognising Ozga's comment about the tendency of experienced policy makers to 'control' the interview, the researchers may have been rather less of the 'attentive audience to the public servant', relying on the 'Themes' and 'Key moments' for points of reference in a desire to 'fracture that polished surface' (Ozga, 2000, p. 127). In this instance, the presence on the team of experienced policy actors and practitioners in ITE proved very valuable indeed.

It was more difficult to gain access to Scottish Executive policy officers and to members of the Inspectorate in Scotland than to secure interviews with DfES officials and HMI/OfSTED officials in England. We also needed to take account of the range of institutional reorganisation which had taken place in England while institutional arrangements had remained more stable in Scotland. So, for example, it was important to get insights into the Council for the Accreditation of Teacher Education (CATE) as well as the Teacher Training Agency (TTA) and to interview pre-OfSTED as well as post-OfSTED HMIs in England.

Comparison of samples (see Table 7) may also reflect the smaller, more cohesive policy community in Scotland, within which more balanced sharing of stakeholder power is achieved between higher education, national organisations and central government, as compared to England. There is also the methodological issue that the research team may feel more secure in generalising from its interviews in Scotland as representative of the stakeholder community, given that the interviews may cover a larger proportion of key stakeholders. This perhaps raises an issue in any comparative project between Scotland and England, where conducting roughly the same number of interviews within the two systems clearly covers very different proportions of the key figures in each system.

Table 7: Overview of the data collected in the historical and policy strands

	SCOTLAND	ENGLAND	TOTAL
Historical	4	7	11
Policy	8	8	16
Total	12	15	27

A second methodological issue arose from the co-existence, at national level in England, of a number of models of ITT. Intra-national conceptual and procedural differences threatened to outnumber Anglo-Scottish differences and with them the legitimacy of the cross-national comparative agenda altogether. As Baistow (2000, p. 10) points out, homogeneity and uniformity rarely exists within a country's borders and therefore when differences are evidenced and noted within national borders, considerations have to be given as to 'how they should be interpreted and at what point *intra*-country variations make inter-country comparisons untenable'. The issue of comparative complexity was compounded by the fact that university departments of education in England and Scotland for instance offer a number of different models or routes which vary in duration and organisation, but that in the original project design it was agreed that each institutional case study would focus on the *overall* provision as opposed to one programme only (for instance postgraduate) or one teaching level only (for instance primary). Finally, an additional unexpected level of complexity surfaced when we realised that it was almost impossible to have directly comparable fieldwork experiences across sites, both for logistical reasons and because processes and procedures are different.

Nevertheless, as will be revealed in the coming chapters, the overall research approach adopted and specific study design have enabled us to gain important insights into the distinctive features of initial teacher education policy and practice in each of the two countries.

Notes

1. In England the term 'initial teacher training' (ITT) is used whereas in Scotland 'initial teacher education' (ITE) is the preferred term. We use each as and when we are referring to one country rather than the other and ITET (initial teacher education *and* training) when we are referring to both or to the wider picture.
2. In Scotland, only two schools opted for self-government while in England hundreds of schools opted out of LEA control.
3. Post-1992 university.
4. Recent policy developments started with 'Mentor Teacher Initiative (1992–5)' in Scotland and with the publication of *Standards for the Award of Qualified Teacher Status, 1998* for England.

Chapter 2

POLICY TRAJECTORIES IN SCOTLAND AND ENGLAND

Introduction

The purpose of this chapter is to provide an account of the policy context in each country in order to set the scene for the subsequent analysis in Chapters 3 and 4 of aspects of current provision in ITET in the two countries. Building upon the account of the broad methodology of the study provided in Chapter 1, we firstly emphasise the importance of an historical perspective in work of this kind. We then consider the nature of education policy making and look in particular at the notion of a national 'policy community'. The development of ITET policy in Scotland and England respectively is then set out, drawing on data gathered during the study, especially from policy makers themselves as well as from providers. The chapter concludes by considering a number of key questions that have emerged and that link to the underlying themes of the study, as set out in Chapter 1.

Why an historical perspective?

This study has at its core an interest in the relationship between policy and practice. In considering practice our focus is on contemporary activities, in some senses we are seeking to provide a 'snapshot' of what was happening in Scottish and English ITET respectively during the years 2003/04. However, in adopting a strongly socio-historical perspective (as outlined in the previous chapter) we are seeking to emphasise the dynamic nature of policy and its continuous interaction with practice. Such an approach seeks to avoid the pitfalls of what Grace (1984) has called a functionalist policy science approach, by contrast with his preferred paradigm of 'critical policy scholarship'. The latter seeks to relate the experience of contemporary actors – for example, student teachers or teacher educators – to a structural, historical and political analysis. It is in other words a critically oriented approach to enquiry. As Grace puts it himself:

> Freed from the inhibition that our thinking in education cannot be allowed to be structural or political then it becomes clear that

the work of critical scholarship, rather than being sterile is, on the contrary, fruitful at an intellectual, professional and practical level. (Grace, 1984, p. 43)

Indeed, drawing on much recent work in education policy sociology, we would further suggest that the notion of 'policy trajectory' is an important one that draws attention to the ways in which policy is best seen as a process rather than as a product or outcome. Bowe and Ball (1992) have described the 'policy cycle', that draws in the processes that influence a policy as efforts are made to implement it, through, for example, the accommodations, subversions or resistances that teachers may bring to bear. Bowe and Ball (1992) also draw attention to the three contexts of policy: the context of influence, the context of text production and the context of practice. Through the multi-method design of this study, it has been our intention to gain insights into all three of these contexts and in the chapters that follow to explore the relationships between them. Referring to the example of the 1988 Education Reform Act in England, they write:

> Education policy is still being generated and implemented both within and around the educational system in ways that have intended and unintended consequences for both education and its surrounding social milieu. (p. 19)

In short, we see policy as a continuing and continuous process, constantly interacting with the material world of teachers, students and policy makers. Education policy and practice are the sites of contestation over values and commitments held by the various actors, both individually and collectively. In teacher education especially we may see some particularly significant manifestations of this, as the various players seek to influence the future shape of teaching and teachers' work within their society. Before beginning to present some of our data about the policy process in the two countries and our analysis of it within the paradigm outlined above, we first summarise what has been suggested in earlier work carried out in Scotland and England, particularly about their respective 'policy communities'.

Policy communities in Scotland and England

The concept of a policy community in fact owes much to work carried out in Scotland during the 1980s. While the concept is now largely accepted as an important one that helps in understanding the policy process, the introduction of the notion of 'community' was powerful in suggesting that there was a sociologically interesting phenomenon that was a key part of the policy process. In essence, this was the identification of a set of shared values among the most powerful policy makers. Moreover, very often this set of shared values was more implicit than explicit. The two key studies in Scotland were by Humes (1986) who described the 'leadership class' in Scottish education and by McPherson and Raab (1988), who suggested that there was an 'assumptive world' shared by those at the heart of policy making.

Humes was scathing in his attack on the extent to which self-interest motivated the actions of the members of the leadership class, accusing them of 'bureaucratic expansionism, professional protectionism and ideological deception' (Humes, 1986, p. 201). McPherson and Raab on the other hand used less impassioned language to describe very much the same phenomenon, the sense of a social selection process which leads to a world view within the policy community that is in some real sense a partial one:

> The assumptive world of the educational policy community was deeply persuasive to those who shared in it. It ordered their understanding of the nature of Scotland and the schools that served it, and it did this by ordering their sense of themselves and the service they could give. Their socialization to a common identity was part of a wider process of sorting and selection by which the policy community was constituted, making the world of practice possible, and with it the reproduction of a social order. (McPherson and Raab, 1988, p. 499)

If these critiques were true of the Scottish education policy community in the 1980s, the extent to which they hold true in the early part of the twenty-first century, at least in relation to ITE, may be seen as a 'sub-plot' of this study.

Studies of education policy making in England on the other hand may be plotted through a series of investigations carried out from the 1970s onwards. At the risk of over-simplification, it may be suggested that there have been three dominant paradigms over that period, influencing how both the policy community and the policy process have been perceived and understood. In the wake of the post-war 'social democratic settlement', sociologists expressed confidence in the democratic process and portrayed the way in which different interests were manifested through the parliamentary process and in education particularly in the partnership that existed between parliament, local education authorities and teachers. So, although there was often a significant conflict of underlying values, this was effectively managed and mediated through due bureaucratic process. This Weberian view of the world is well demonstrated in the work of Maurice Kogan (1975).

No doubt the policy community of this era could have been subjected to much the same sort of critique as that applied in Scotland, described above. However, the dominant critique that emerged within educational sociology in England, and challenged the partnership paradigm, was one informed by Marxism and by the increasing concern that social democracy, including the education system, was failing to challenge the structural inequalities within society. In this view, education policy was very much determined by the central state. The clearest and most developed expositions of this view were set out by Roger Dale (1989).

The emphasis on 'structure' within this paradigm was seen by some to be over-deterministic. In an attempt to bring 'agency' back in, including that of teachers, what Stephen Ball (1994) described as a 'critical and post-structuralist' approach emerged, that reasserted contestation and processes of influence, while still acknowledging the central role of the state, in a range of forms. The current manifestation of this paradigm however is challenged by the need to take fully into account the 'Third Way' of New Labour, with its adoption of a mixed economy of provision where the boundaries between the public and private sectors have become porous and sometimes invisible (see Fielding, 2001; Newman, 2001; Crouch, 2003).

There is no doubt that the policy contexts in Scotland and England have significant differences both historically and contemporarily. However, the concept of a policy community is very important in both contexts and some of the current trends, such as those associated with New Labour, may be identified in both settings, as we shall see. We turn now to examine the particular trajectories of policy regarding ITET in Scotland and England respectively.

The development of initial teacher education policy in Scotland

From monotechnic colleges of education to university merger

Although under pressure to rationalise during the period of severe student intake contraction in the late 1970s and early 1980s, ITE in Scotland continued to be provided almost entirely in monotechnic colleges of education throughout the 1980s, with the single exception of the University of Stirling which had introduced an ITE programme of a distinctive type from its establishment in the 1960s. Indeed, formal review of the college of education sector carried out during the 1980s by the Scottish Tertiary Education Advisory Council (STEAC) in 1985, seemed to reinforce the general view of stakeholders that ITE should continue to be provided in monotechnic institutions (see Kirk, 2000, pp. 3–5).

Senior figures within the Scottish college of education community at that time certainly seemed to have appreciated the greater control over professional activity involved in monotechnic status. As one former Principal of a college of education whom we interviewed commented:

> I felt that if we were to become part of the universities, we would lose control of what we are doing. If we kept our institutional identity,

we could control what happened. For me the essential thing was the links with the profession, being part of the professional community.

This former Principal of a college of education also highlighted the attraction to the Scottish Office of the monotechnic status:

> Scottish teacher education was conducted in institutions that were separate from universities by a process of deliberate, calculated policy because the people who ran the Scottish Education Department had no confidence in the university to deliver teacher education. Scottish teacher education had to take place in institutions that they could control.

However, within a ten-year period beginning with the merger of Jordanhill College of Education and the University of Strathclyde in 1992, all the former Scottish colleges of education completed merger with universities to become Faculties/Schools of Education fully integrated into the university sector. The background to, and reasons for, this process of merger have been analysed fully elsewhere (see Kirk, 2000, pp. 5–15). However, for current purposes, it is interesting to comment on the staff perceptions within the now university Faculties/Schools of Education of some of the impacts of this merger process. All staff recognise that their professional lives have changed fundamentally. As one Head of a Faculty/School of Education commented:

> Now we are part of the universities, obviously we have to fit in and respond to university agendas as well, and as the new boy as it were in the university you have to make a pretty strong case if you are going to do things differently from the university.

This has required Heads of Faculties/Schools to seek significant change in the professional practices of their staff. As one Head of Faculty/School commented:

> I am trying to encourage them to look really critically at how much time they spend teaching. Looking at the kind of hours that an undergraduate say doing history would spend being taught directly and comparing that to a BEd student, BEd students get much more teaching, does that mean they are learning more or less? These are the kinds of questions that I am encouraging people to think about.

Other staff have highlighted the particular pressures in meeting university research expectations at the same time as carrying out other student-focused and partnership activities:

> You get no acknowledgement for the hours you spend organising and trying to get tutors or students to do things, whereas if you write a research paper there is kudos in that.

Heads of Faculties/Schools can be seen responding in different ways to these tensions. For example, one Head of Faculty/School described how his university was recognising the pressures on staff trying to combine professional teaching and research roles by appointing two separate categories of staff to permanent contracts:

> The idea that there should be university teacher and senior university teacher roles, when we put out advertisements now we may go for university lecturer, senior lecturer or we may go for a university teacher or senior university teacher.[1]

Another Head of Faculty/School was deeply unsympathetic to such an approach, asserting instead his strong belief that all permanent lecturing staff should be active researchers:

> The conclusion that you didn't have to be a researcher to have research-based teaching, I am pretty sceptical about that. You sit there and you read it but you don't do it. I am dubious about that, I don't regard that as an adequate claim to expertise.

The governance of Scottish ITE

Certainly, by becoming Faculties/Schools of Education within Scottish universities, the ITE providers have been required to work within the general Quality Assurance frameworks of the university system. For example, the Standard for Initial Teacher Education was produced in 2000 by a Joint Working Party including representatives from the Quality Assurance Agency for Higher Education (QAA) along with the more traditional stakeholders within Scottish initial teacher education, that is, staff from the Faculties/Schools of Education, the GTCS, HMIE, local authorities and schools. Of course, the presence of QAA and university representatives within this stakeholder group does confirm that the university providers of ITE have a secure role in the joint production of the national Standard governing ITE. This Standard for Initial Teacher Education (SITE) was to be linked to a 'one-stop' Collaborative Review process for ITE, which would include all stakeholders involved in the production of the SITE itself (i.e. the intention was to merge into a single exercise the separate quality assurance visits which GTCS, HMIE and higher education agencies had previously conducted). In the end, this Collaborative Review process was not implemented because of GTCS resistance to a one-stop approach which removed the GTCS's separate entitlement to review ITE programmes. Crucially however, SITE itself remains the governing Standard for Scottish ITE, which indicates a consolidated role for Higher Education in contributing to the governance framework for ITE.

The fact that SITE was produced by a Joint Working Group of a range of stakeholders which did not actually include Scottish Executive Education

Department (SEED) officials themselves provides a recent illustration of the general point that the SEED influence on ITE in Scotland tends to be exercised indirectly through other key stakeholders, rather than by SEED itself. Certainly, SITE incorporates the most recent version of the earlier Competences for ITE, which were produced directly by SEED's Scottish Office predecessors. There had been Guidelines for ITE programmes produced by the Scottish Office in the 1980s. However, the key development is often seen as the Guidelines produced in 1993, which incorporated explicit Competences, the inclusion of which was much commented upon at the time (see Carr, 1993; Humes, 1995; Stronach et al., 1994; 1996). There was significant academic concern expressed from some quarters within Scottish ITE when these Competences were first introduced, and there is evidence that such concerns still remain. For example, one Head of Faculty/School expressed very strong hostility to the requirement to work within such approaches:

> What's happening at the moment is that the agenda has been dominated by the kind of standards/competences, that kind of rational, pseudo-rational analytical framework, because a) it has got a tremendous superficial attraction b) it provides a mechanism of control.

Other ITE staff seem less concerned with any 'threat' posed by such Competences to the freedom of HEI providers to construct and deliver appropriate ITE courses. For example, a second Head of Faculty/School commented:

> The competences blend the subject specific with the generic classroom skills, with the cross-curricular and whole-school dimension, with the broader professional skills and commitments around reflective practice in the deepest sense, and with the reading of academic literature and the need to gain an appropriate level of research skills. These give a sufficiently broad framework within which you can take the students forward in all kinds of directions.

Of course, as has already been mentioned, the Guidelines, with their associated Competences, have in a sense been overtaken by SITE. Beyond this, other examples of SEED's indirect exercising of control can be provided. These include the Scottish approach to ITE funding. SEED advises the Scottish Funding Council (SFC), which then allocates funding for ITE to university providers. Of course, while ITE recruitment places are allocated formally to the universities by SFC, these allocations are tightly based on the SEED advice to SFC. The role of HMIE within Scottish ITE also demonstrates indirect SEED influence. Formally, HMIE is no longer part of the SEED *per se*. Rather, HMIE has semi-independent agency status. Additionally, HMIE now only makes 'Aspect Reviews' at the request of the Scottish Executive of particular themes within ITE (such as Literacy Across the Curriculum or Student Placements). From such Aspect Reviews, HMIE only produces sector-wide reports, rather than institution-specific evaluative reports of the performance of particular HEI providers. Finally, as illustration of SEED's indirect role, SEED has delegated its approval of ITE programmes to the GTCS.

The position of the GTCS in relation to ITE raises interesting issues within Scotland. Clearly, the presence of a strong General Teaching Council in Scotland since the mid-1960s can be presented as evidence of significant professional control within the Scottish context for schoolteaching, especially in contrast to England. As already indicated, the GTCS has had longstanding powers of accreditation and review of ITE programmes delegated to it by Scottish administrations. These powers give elected teachers serving on the GTCS an influence on the development of ITE. However, there is very considerable evidence of unease among other key stakeholders about the role of the GTCS in relation to ITE. HEI staff often comment upon the restricting influence which they see in the GTCS, preventing the development of more innovative approaches to ITE. For example, one Head of Faculty/School commented:

> Somebody's teaching in a school one day and the next day they are called out to do an accreditation panel. They don't know anything about ITE except their own experience. What are they going to do? In a sense they have to follow what the guidelines say. They are, I suppose, to be guided by professional officers of the GTCS to a certain extent, so it does depend partly on what the professional officers think and if you get bureaucrats then you get bureaucracy. And bureaucracy is just a nuisance.

Other stakeholders have been even more outspoken. For example, a senior Director of Education with a particular interest in ITE commented:

> I find the GTCS at Council level most frustrating. There is a basic problem which the Scottish Executive has got to face up to in the GTCS. On the one hand it is by and large self funded – it is paid for by the teaching profession, therefore it is run by the teaching profession. Can we actually afford as a Scottish Executive to have a watchdog on the teaching workforce by and large in the hands of a union? And I think the answer is no. It has a vested interest and I think it has actually become an impediment to the interesting developmental work we are doing.

In his comments on the GTCS, this Director also raises the specific issue of professional association/trade union power within any stakeholder analysis of Scottish ITE. In Scotland, the Educational Institute of Scotland (EIS) dominates the scene. With a total membership of 53,000, it represents almost all Scotland's primary teachers, the majority of its secondary teachers, and substantial numbers in further and higher education. The EIS works systematically to ensure that EIS-sponsored candidates are elected to the GTCS, and secure dominant positions within the Council. Views differ on the impact of this EIS role. Many other stakeholders and commentators will present this as a negative, conservative influence on areas like ITE, with EIS figures tending to use workload and resource concerns persistently as barriers to innovation and change. Others will argue that the EIS genuinely attempts to take the broader, more progressive role identified in its founding Royal Charter of 1847 'for the purpose of promoting sound learning and of advancing the interests of education in Scotland'. Such commentators will point to recent EIS initiatives like its partnership with a university (the University of Paisley) as a provider of a Chartered Teacher programme.

Highlighting the views of a senior local authority Director also interestingly raises the broader issue of the role of Scottish local authorities as stakeholders within ITE. Certainly, in contrast with England, ITE in Scotland has always been provided by higher education institutions working in partnership with local authorities and schools. However, it can also be suggested that the direct involvement of local authorities themselves in these partnerships has tended to be very limited to date. For example, again a senior Director of Education has commented that:

> The supply of teachers is crucial to the work of the local education authorities, the quality of teachers is crucial to that and yet I am regularly disappointed in the involvement that we actually have. Also I would probably, talking from the local authority point of view, say there is a lack of urgency on our part at times to actually become involved and we almost stand back and see the teacher education institutions and the schools making all of the arrangements.

However, there have been significant moves since 2004 to increase significantly the formal role of local authorities within ITE partnership. Issues of partnership within Scottish ITE will be explored more fully in Chapter 4. At this point, it can perhaps be noted that the increased role of local authorities since 2004 has focused more on the operational involvement of newly identified local authority placement co-ordinators in order simply to secure placements for the very significantly increased ITE intakes required at postgraduate level from 2004 onwards. In a sense, this increased local authority role can be seen as a response to the 'near panic' concern that significant numbers of these greatly increased ITE intakes would remain unplaced for school experience.

ITE in Scotland post-devolution

These comments on the role of local authorities lead to more general consideration of the political environment for ITE in Scotland in the post-devolution context. In considering the impact of Scottish devolution upon any aspect of the Scottish education system, including ITE, one of the challenges for analysis is that the devolution settlement coincided with the establishment of a Labour government at UK level after its 1997 UK general election success. Therefore, the challenge for analysts is to distinguish between the impact of a new UK Labour government upon Scottish affairs generally, and the more particular impact which a specific Edinburgh-based Labour/ Liberal Democrat Executive may have had. Certainly, as Chapter 4 will explore, the replacement of a very unpopular Conservative administration in Scotland from 1997 could have been expected to unblock a number of the policy impasses which that Conservative administration had experienced in Scotland. Apart from more general educational policies, such as Scottish resistance to self-governing schools, ITE in the 1990s demonstrated specific Scottish resistance to the Conservative administration's Mentor Teacher Initiative and its attempt to move towards a more school-focused approach to ITE, on the model being developed in England at the time.

Certainly, the new post-devolution Scottish administration commissioned a First Stage Review of ITE conducted on its behalf by Deloitte & Touche (Deloitte & Touche, 2001). This led to the production of a specific Action Plan by SEED for implementation of ITE developments from 2001 (Scottish Executive, 2001b). Subsequently, the Scottish Executive undertook a Second Stage Review of Initial Teacher Education between 2003 and 2005 (Scottish Executive, 2005a; 2005b).

However, arguably these reviews of ITE have not led to policy initiatives as forceful as those which the post-devolution Labour/Liberal Democrat administration has taken forward for other aspects of educational policy relating to schools. The strongest example of a major Scottish Executive initiative in relation to schools and schoolteachers has probably been the McCrone settlement. Following the McCrone Report of the Committee of Inquiry into Professional Conditions of Service of Teachers (Scottish Executive, 2000), the Scottish Executive reached the agreement *A Teaching Profession for the 21st Century* (Scottish Executive, 2001a). This settlement involved major salary increases for Scottish schoolteachers, and associated reforms in the conditions of employment for Scottish schoolteachers, e.g. the introduction of a guaranteed one-year induction post for all qualified Scottish ITE graduates, enabling them to achieve the new Standard for Full Registration within one year, as opposed to the previous two-year minimum probationary period. It can be argued that this very 'pro-teacher' national settlement on teachers' pay and conditions was significantly in advance of the approaches to teachers' pay and conditions being taken for England and Wales by the UK Labour government at Westminster.

Similarly, another distinctive Scottish Executive initiative since 2004 has been the very significant increase of student intakes onto Scottish ITE programmes, especially the PGCE/PGDE² programmes. This can be associated with the Scottish Executive commitment to produce a total teaching force in Scottish schools of 53,000, especially linked to commitments to reduce class sizes in the early years of primary school, and in English and Mathematics in the early years of secondary school. Certainly, this particular initiative impacted upon ITE. However, perhaps ironically, it can be argued that the initiative to increase student teacher intakes indicates the comparative difficulty of raising the internal issues of ITE innovation and development sufficiently high up the political agenda. In contrast to the very specific resource commitment to increased ITE intakes as a means of achieving a broader political objective for schools (i.e. reduced class sizes through increased teacher numbers), the underlying resource issues for supporting progressive development of partnership in ITE have not been similarly addressed. Indeed, it can be argued that the Scottish ITE system, in responding very positively and successfully to the Scottish Executive's request to increase ITE intakes, has placed itself under such severe pressure simply to deliver existing forms of programmes to these increased intakes, that potential for genuine innovation has in fact been constrained.

Absence of supply pressures on ITE in Scotland
The success of the recent recruitment campaign to secure very significantly increased ITE intakes illustrates an important contextual point about Scottish ITE. Certainly since the late-1970s, Scotland has not faced the type of severe challenges of teacher supply which have been encountered in England (Menter, 2002). Even if particular secondary subjects have at times been more difficult to recruit than others, intake targets have been met. For example, when commenting that Scotland had not considered alternative school-based training routes such as those developed in England, a former senior GTCS official emphasised:

> Just because certain things were being developed south of Hadrian's Wall it didn't necessarily mean that we ought to be doing it. Of course all this was coloured by the lack of shortage and although I was critical of what was happening south of the border it was understandable. I mean they were horrendously short, particularly the inner cities and had to find all kinds of ways of encouraging people to come into teaching. Now, we didn't have that.

The development of initial teacher training policy in England
Central government intervention in ITT
The somewhat turbulent recent history of initial teacher training in England has been well documented in a number of places (e.g. Gilroy, 1992; Furlong *et al.*, 2000). In the period before the mid-1980s, initial teacher education had

increasingly been located within mainstream higher education, including a large number of university and polytechnic departments of education, as well as colleges of education (Dent, 1977). Since the mid 1980s there has been a series of central government interventions which have gradually increased government control of the processes by which one may become a teacher.

1984 saw the establishment of the Council for the Accreditation of Teacher Education (CATE), the first such national body in England (Circular 3/84: DES, 1984). Criteria for course accreditation were introduced, setting out a number of requirements that providers had to meet, as well as some stipulations about the structure and content of courses for primary and secondary courses.

There followed a series of government circulars (24/89: DES, 1989; 9/92: DfE, 1992; 14/93: DfE, 1993; 10/97: DfEE, 1997) which by 1998 (Circular 4/98: DfEE, 1998a) had led among other things to the creation of a national curriculum for ITT and the definition of a series of standards, the achievement of which was necessary in order to qualify as a teacher (Mahony and Hextall, 2000). From 1992, there was a major drive by the government to increase the role of schools and school staff in initial teacher training, at the expense of HEI involvement. The period was also characterised by the active participation in these processes by two key bodies: the Teacher Training Agency,[3] established in 1994 (and replacing CATE), to fund and approve provision; and, the inspection agency, OfSTED, established in 1992, which has played a major part in assessing the extent to which particular courses or institutions were providing high quality.

Diversification of routes for ITT

There has also been increasing diversification in the nature of pre-service provision, with school-centred and employment-based routes being added to the more traditional HE-led programmes such as Bachelor of Education (BEd) and Post Graduate Certificate in Education (PGCE) courses. But even these more traditional routes are varied now, with BEd courses of from two to four years in duration, PGCE courses which may be 'flexible' and/or part-time and may be one or two years long. SCITTs were established in 1993 and enable consortia of schools to set up their own training scheme leading to QTS, which may or may not involve an HEI. The employment-based routes established in October 1997 replaced the articled and licensed teacher routes (see OfSTED, 2002) and enabled schools to employ trainees while they worked towards their qualification. In summary, although the traditional routes still provide the majority of new teachers in England, there is now a host of routes of entry to the teaching profession, which have various levels of participation and involvement from different stakeholders, including some approaches that have no HE involvement at all.

In interviewing a range of stakeholders from the policy community about the rationale for the increasing diversification of routes of entry we have identified a number of strands. First, there is a suggestion, most frequently coming from HEIs themselves, that the key motivation was ideological. It was another example of a New Right attack against 'producer capture'. Evidence to support this view commonly comes from citing pamphlets of the 1980s and early 1990s, as well as from politicians' speeches, such as Kenneth Clarke's North of England Conference speech in 1992, where he accused providers of promoting 'barmy theory'.

For example, we find a GTCE officer commenting:

> Political considerations came into it, the government had a downer on HE, that it was full of left-wing ideologies and people who couldn't teach for toffee. Therefore let's get it [training] into schools and you know the model you have of teaching is such that if you just sit around and teach long enough, you will become one.

Similarly, a senior HEI manager commented:

> I don't think there was any doubt the TTA did want to undermine teacher education institutions. I think there is no doubt they wanted to erode public confidence in the way they [teachers] were trained and look for other roads. Anthea[4] used to say 'you do not need buildings to train teachers all you need is schools and some willing teachers' so the articled teacher scheme came along in the belief that it was much more school-based, the students would understand and know much more about what was expected and retention would be better.

The second strand is a technical rational type of argument, which is sometimes acknowledged by the providers but more frequently used by civil servants and by TTA officials. This line suggests that there were problems of quality in the teaching workforce and that in order to address this, the training system needed to be 'fixed' or 'sorted out' in some way. Opening training up to greater involvement of schools was seen as part of the solution, as indeed was the original creation of the Teacher Training Agency.

As one DfES official commented:

> Looking at the diverse routes into teaching, we needed [to be] far more flexible. If I look at the student associate scheme that was developed and that very much came out of a think tank idea . . . it was part of a white paper commitment where officials and some of the advisers were to come up with something that could encourage good, perhaps better quality, graduates to consider teaching. But you wouldn't do that in isolation, you would talk to people at the TTA and perhaps people at the TTA were very much at the forefront of developing the detail.

On the teachers produced from such schemes, this official commented:

> I think it depends whether you think that they do come out with a different sort of professionalism, but I think what we are saying here is that it is very much outcome-focused so at the end of the day these are the standards that you have to achieve and this covers the professional values in practice.

The third rationale which came increasingly into play during the 1990s was about supply, recruitment and retention. This is variously expressed as solving an impending crisis in supply, or ensuring that there is a more representative teaching workforce, or getting at good quality people who would not consider coming into teaching through the conventional routes.

Some of these views were summarised by a GTCE officer:

> We think there are a number of good arguments for having diverse routes into the teaching profession. One is that when we have had more monolithic routes, we haven't attracted a good cross-section of society and, if that is linked to the routes, then we need to think about other ways of doing that. People have different circumstances and, particularly for the career changers, it is desirable to make sure that we have got ways of becoming a teacher that are accessible to the widest range of people who are suitable, and interested.

Quality issues with new ITT routes

However, there are clear concerns about the quality of these new forms of provisions.

One HMI commented:

> It is quite a tough thing for a government to accept, but I think it may be that inspection evidence is showing that the problems of flexibility, the problems of distance mean that there will almost inevitably be a loss of quality unless you rack up the unit of resource hugely.

Following the political criticism of teacher education in the 1980s and early 1990s, it is perhaps not surprising that many of the traditional providers expressed grave concern about the impact of the new approaches to the quality of the provision:

A senior HEI manager emphasised:

> They can get QTS if they have met the standards, they get QTS and that's it. I am very dubious about those school-based routes that involve very little HE. I mean I just think that it is teacher training on the cheap. Actually it is very expensive for the TTA who pay them (trainee teachers) a salary, but I think it's a cheap way of educating them because I think they miss out.

Further, it must be noted that early inspection reports did much to confirm these anxieties. There was growing concern, including among members of the inspectorate, that the new schemes might be producing teachers who were only 'adequate' and who might not be equipped with the appropriate foundations for continuing professional development.

In interviews, one HMI commented emphatically on employment-based routes:

> [We have inspected] SCITTs, but on the employment-based route we had done the old registered/articled teachers stuff and that worried the life out of us at the time. So, we did a survey of the GTP and reported one of the most critical reports probably that we have produced for a long time. We fed that back in some detail to the TTA and the DfES sometime before we published, so that in a sense they could get their retaliation in at the same time. As a result of that critical report there was an immediate very thorough review of GTP leading to a re-launch with designated recommended bodies and all the involvement of proper accredited providers, proper quality assurance requirements and so on. Of course GTP is now growing into something of a very major player in ITT in England.

This HMI had similar comments to make on SCITTs:

> The reports that we published in the summer spelt out much more clearly than any previous report that I have seen from OfSTED that provision in SCITTs was as a whole consistently weaker than in HEI-based partnerships. That message has been taken to heart very seriously by the TTA, they have established a team within the TTA to support SCITTs, to help SCITTs to ensure that their provision matches the quality in HE providers. However, there are some issues which are generic to school-based training.

This HMI's comments provide a very useful overview of the HMI evaluation of employment-based routes and SCITTs, and it seems important to summarise this overview to stand as more general comment on these routes. HMI stated their review of the GTP (employment-based route) (see OfSTED, 2002) was their most critical report in a long time, and HMI reported that provision in SCITTs was as a whole consistently weaker than in HEI-based partnerships. In both cases this led to a remedial plan by the TTA which was very committed to these routes. For example, specific support was put in place for SCITTs. GTPs have been re-launched with designated recommended bodies and involvement of properly accredited providers and quality assurance requirements. As a result, the GTP route has not been abandoned, but re-launched and is now growing into a very major player within English ITT. However, HMI still considers that there are issues of quality which are generic to school-based training. Indeed, our HMI interviewee conceded that routes which raise these concerns about quality can really only be justified while there are still teacher shortages:

> The one thing I would add in terms of policy is that one of the real battles that we are going to have is over quality because of the issue of teacher supply and retention. It may be the situation that alternative routes, like GTP, will generally at best deliver adequately trained teachers. Actually, the cost of doing more than that just simply would make it no longer cost effective. However, as long as they produce teachers who meet the standards, then may be that is all that they can do for most. Now the question then is, is that acceptable? If GTP, for example, is providing real additionality in terms of supply of mathematics teachers, science teachers, modern foreign language teachers, then it may be that we have to accept, quite publicly, that they won't be as well trained on average as if they go to the University of [X] or wherever, but they are adequately trained and they are trained where they wouldn't otherwise be.

The current English position on ITT – some concluding comments

If the motivations for policy developments in diversifying routes did stem from this range of supply concerns, then the new situation that has developed, in which the major concerns about supply are significantly reduced and indeed are limited to relatively few geographical areas and subjects, creates an interesting question. Will there be a reversion to the traditional modes of entry, that were so vociferously if ineffectively defended by the HEI providers and, albeit to a lesser extent, by teacher unions? Of course, the nature of the policy community in England is very different from that in Scotland. The GTCE was not established until 2001 and has thus far had extremely limited powers in respect to ITT (although greater influence on Continual Professional Development (CPD)). There is a greater variety of unions, with none of them being anything like as dominant as the EIS is in Scotland. The seven ITE providers in Scotland work closely together through the Scottish Teacher Education Committee (STEC) and through this have direct collective contact with the other stakeholders. In England there are now effectively hundreds of providers of ITT. Some of these do have collective bodies such as the Universities Council for the Education of Teachers (UCET), the Standing Conference of Principals (SCOP) and the Association of SCITTs. However, the different categories of providers do not have a united voice with which to speak to government or its agencies. Moreover, these relative weaknesses or absences in countervailing stakeholder strength in England have to be set alongside the very significant power and influence of the TTA and OfSTED, with the DfES effectively operating the strings, at least for the TTA. All of this would suggest limited prospects for a return to the exclusive use of ITT routes involving HEIs.

Policy trajectories in Scotland and England: discussion and conclusions

Having reviewed the policy contexts for ITET in both Scotland and England, a number of interesting questions emerge, relating to our underlying theme of convergence or divergence between the two systems. We can attempt some preliminary conclusions on the relationship between Scottish and English developments and the wider European and global trends in ITET suggested in Chapter 1. In particular, we can suggest preliminary judgements on the extent to which Scotland and England converge or diverge in relation to these European and global trends.

In terms of the European and global trends around public sector accountability, performativity and new managerialism, there does appear to be much more explicit evidence of these approaches in England through the direct roles exercised by the TTA and OfSTED. These appear to contrast with the less direct approaches taken in Scotland, such as the Collaborative Review proposals and the more conditional role for HMIE. On the other hand, there may be less divergence between Scotland and England on this issue than at first appears. The universitisation of ITE in Scotland can arguably be seen as part of the public sector accountability process. It can be suggested that policy makers in Scotland were more relaxed about moving from their direct control mechanisms over Scottish monotechnics and into the university system because simultaneously a more formal national quality assurance framework had been established for the universities, with SHEFC TQAs, the role of QAA in producing SITE and the proposed role for QAA in the Collaborative Review of ITE.

The mergers in Scotland of the former economically expensive monotechnic colleges of education into the university sector from the early 1990s can be seen as part of the wider European and global trend of rationalisation in ITE provision. This may indicate a divergence between Scotland and England in the period from the late 1980s, given that there had been earlier changes in England prior to this period in the relationship between teacher education provision and the wider higher and tertiary education sector.

The universitisation of Scottish ITE from the early 1990s can also be seen as part of the wider European and global trend of professionalisation of teacher education and the teaching profession, in particular as part of the upgrading of academic status associated with professionalisation. This may appear to suggest an important divergence from England, where arguably the most significant alternative trend in ITT provision was the development of the 'apprenticeship model' involved in GTP and SCITT approaches, with an apparent associated downgrading of higher education academic involvement in the ITET process. Of course, the alternative point can be made that the university sector in England had already established a greater role in ITET than was the case in Scotland prior to the late 1980s. We will revisit this issue in Chapter 3, where consideration of the respective standards for ITET in the two countries suggests that there may be clearer statements in Scotland than in England about the current importance of the theoretical and research underpinning of teaching within ITET standards.

It may also be suggested that there is significant divergence between the contexts for ITET in Scotland and England because of the different approaches to the professionalisation of schoolteaching within the two systems. For example, it can be argued that the Scottish Executive since devolution has been much more 'pro teacher' than the Westminster Labour government. The very positive pay and conditions outcomes awarded to Scottish schoolteachers in the McCrone settlement can be suggested as evidence for this. This can be taken as an illustration of significant divergence between the two countries on the wider trend of professionalisation of schoolteaching. The recent policies of the Scottish Executive could be described as reflecting the longer-standing, relatively higher status of the schoolteaching profession in Scotland as compared to England. This higher status can be presented as the result of a number of longer-term economic, cultural and historical factors. Traditionally, this more established professionalisation and higher status of the schoolteaching profession in Scotland can also be equated with stronger direct political influence, especially through its dominant professional association/trade union, the EIS. It may be suggested that this stronger professional status of the Scottish schoolteaching profession explains the divergence between Scotland and England, where Scotland has not adopted the significant increase in non-university ITET developed in England. We will return to these themes in Chapter 5.

Notes

1. The distinction here being that university teachers/senior university teachers have no contractual requirement to complete research activity.
2. The one-year postgraduate primary and secondary ITE programmes in Scotland are replacing their previous title PGCE (i.e. Post Graduate Certificate in Education) with the new title PGDE (i.e. Professional Graduate Diploma in Education).
3. In 2005 the TTA became the TDA, the Training and Development Agency for Schools, but we refer to it throughout this book as the TTA, since that is what it was called during the time of our study.
4. This is a reference to Anthea Millett, the first Chief Executive of the TTA.

Chapter 3

FROM POLICY TO PRACTICE?
THE QUESTION OF PROFESSIONAL KNOWLEDGE

Introduction

In this chapter we are concerned with the question of professional knowledge, firstly interrogating some of our documentary data in order to assess similarities and differences in the ITET 'policy rhetoric', that is, the ways in which the work and role of teachers is officially defined in the two countries (Menter *et al.*, 2006). We examine key documents that set out the framework by which people may qualify as a teacher in each country.

The second stage of the analysis then examines the discourse employed by some of the key stakeholders involved in the formation and implementation of teacher education in each country. In particular we draw from our interviews with teacher educators. This enables us to present an account of how teaching is understood by those with the key responsibility for bringing new teachers into the profession.

Our concern is to identify the underlying values which define teaching in each country in order to consider the extent to which teaching is understood as the same activity. There is considerable evidence that the restructuring of teaching is a global phenomenon (Smyth *et al.*, 2000; Morrow and Torres, 2000; Robertson, 2000) and it is our aim here to gain insights into British approaches to restructuring through seeking to identify what professional knowledge is deemed to be essential in these two countries.

There are other recent perspectives introducing further complexities in interpreting the relative influence of national and global factors on the development of education systems. For example, it has been suggested that education systems ought to become increasingly responsive to local contexts and needs, as the world becomes increasingly 'balkanised'. Arguing along these lines, Parker (1997) suggests:

> Post-modern teachers will construct pedagogy out of local interests and concerns where worth and value is set within a narrative in which its players have a stake and a voice. (pp. 151–2)

Similarly, learning theorists suggest that teachers will need an increasingly sophisticated and research-informed understanding of how young people learn in order to operate effectively in a knowledge-based society in which electronic communication makes instant transfer possible (Edwards *et al.*, 2002); and where learning is recognised as 'situated' and may take place most effectively within 'communities of practice' (Lave and Wenger, 1991; Wenger, 2000).

What is professional knowledge?

The *Concise Oxford Dictionary* suggests one meaning of knowledge is 'theoretical or practical understanding'. If we add to that the prefix 'professional', we can have a simple enough view of the phrase as meaning 'theoretical or practical understanding of teaching'. Of course, within that definition we have already incorporated one of the key problematics in debates about professional knowledge, which is whether the emphasis should be on theory or practice, or indeed what the relationship between the two is.

Eraut suggests that professional knowledge is: 'the knowledge possessed by professionals which enables them to perform professional tasks, roles and duties with quality' (cited by Day, 1999, p. 53). Schon, who has been so influential in the development of 'reflective practice' as a guiding concept in professional education, associates professional knowledge with the technical rationalist view of professional activity. Referring to professional Schools (i.e. departments) within universities, he says:

> The normative curriculum of the schools rests . . . on an underlying view of professional knowledge as the application of science to instrumental problems . . . The schools' view of professional knowledge is a traditional view of knowledge as privileged information or expertise. (Schon, 1987, p. 309)

Schon then goes on to argue that such a limited view of professional knowledge is inadequate to describe the full range of knowledge deployed by successful professional practitioners.

Campbell *et al.* (2004) develop Schon's ideas to relate them to contemporary conceptions of practitioner researchers:

> Part of what it is to be a good practitioner is to be able to bring . . . tacit knowledge ['knowledge as revealed by actions'] to the surface by a process called reflection-in-action, by thinking through one's actions as one is producing them in the thick of one's professional situation. (p. 10)

The most sustained work on the topic within the UK has been by Eraut (1994). He identifies different components of professional knowledge, including *propositional knowledge, personal knowledge* and *process knowledge,* all of which are significant.

Also writing recently, Goodson (2003) argues that professional knowledge is intimately bound up with professional identity. He argues that current debates about its nature reflect a wider set of major changes which are going on in teachers' work:

> The forms of knowledge which teachers have are substantially implicated in the kind of people teachers are and believe themselves to be. At the moment . . . those forms of knowledge are being substantially restructured and, as a result, are substantially changing the kind of people that teachers are and are seen to be. (p. 4)

These are but a few examples from published work that seem to underline the real and symbolic significance of professional knowledge in understanding teaching, teacher education, teachers' work and educational change.

'Professionalism' is a notoriously difficult term, not least within teaching. It carries with it so much of what has been contested about teaching over the past 150 years, with questions of status, pay and conditions, state control or autonomy at various times being fiercely argued. Rather than explore that territory here, let us simply acknowledge that 'professional knowledge' is likely to carry some of these tensions within it, and may be a key element in defining what it is to be a teacher, but also how it is that the teaching profession defends itself from those who argue that qualifications may not be essential.

If we consider the way in which the teacher education curriculum is commonly constructed – at least in England and Scotland – with the three major strands of subject knowledge, (subject) pedagogy and educational studies, which are delivered in two main contexts, the academy and the school, what can we say about where and how professional knowledge is offered to students, how it is learned or developed and who plays the key roles in that?

Key documents

We now turn to the key policy documents which set out the statutory requirements for courses of ITET as well as a governmental vision of the newly qualified teacher (NQT) in each country. In Scotland, the key documents are the *Guidelines for Initial Teacher Education Courses in Scotland* (SOEID, 1998) and *The Standard for Initial Teacher Education in Scotland* (QAA, 2000). In England the key document is *Qualifying to Teach* (TTA and DfES, 2002), first published in 2002 following a much fuller consultation process than had previously been undertaken by the TTA (TTA, 2004a). This is augmented by a lengthy handbook of guidance which was updated in Spring 2004 (TTA, 2004b).

In our analysis of these documents, we focus on how the work of a teacher appears to be understood. In this chapter, we take these documents to best represent the official or state view of what a teacher is or should be.

Scotland

The 1998 document from the Scottish Office sets out a list of competences which beginning teachers must acquire. However at the start of the document there is a very clear statement of priorities for ITE:

> The overall aim of courses of initial teacher education is to prepare students to become competent and thoughtful practitioners, who are committed to providing high quality teaching for all pupils. [...] through teaching and by example, they must be able to foster their pupils' personal, social, emotional and moral development and encourage them to take pleasure in learning. The courses should assist students to reflect on their practice and its impact on pupils and assist them to consider the ways of improving their effectiveness as teachers. They should be informed about how to access and apply relevant findings from educational research. (SOED, 1998, para. 2)

Detailed requirements for each qualification (BEd, PGCE, primary, secondary, etc.) are then set out before the competences which courses 'must enable student teachers to acquire' are listed. These are listed under the following headings:

1. Competences relating to subject and content of teaching
2. Competences relating to the classroom
   2.1 Communication and approaches to teaching and learning
   2.2 Class organisation and management
   2.3 Assessment
3. Competences relating to the school and the education system
4. The values, attributes and abilities integral to professionalism.

However, as indicated above, this government document was soon to be supplemented by the Standard (or benchmark statement) produced by a Working Group consisting of seven HEI representatives (one of whom chaired the group), two GTCS representatives, one person from an education authority, two head teachers, one HMI and an observer from the QAA. Furthermore, it is worth noting again (as described in Chapter 2) that quality assurance relating to the Standard was to be achieved through a process of 'Collaborative Review' involving a similar range of stakeholders, rather than through inspection, as by OfSTED in England.

While a benchmark statement for Education Studies had been prepared earlier by the QAA (under the same chairperson and which applies to all undergraduate courses in this field across the whole of the UK), no equivalent benchmark for ITE was developed in England or elsewhere in the UK. This is thus a distinctively Scottish development. Indeed it could be seen in policy terms as a very strategic move to retain significant control of ITE within HE in Scotland, by contrast with the developments in England.

Certainly the introduction and other parts of the document take great care to position the statement in relation to the earlier competences document (SOED, 1998):

> Attention has been paid to the national requirements for ITE and the document incorporates the competences in the *Guidelines for Initial Teacher Education Courses in Scotland* . . . (QAA, 2000, Introduction)

However the explicit achievement of this document is to take the concept of the benchmark statement (from QAA) and to develop it into the first of what was to become a series of standards for teaching in Scotland, namely the Standard for Initial Teacher Education (SITE).

> The benchmark information is based on a vision of the newly qualified teacher who, having successfully completed a programme in Initial Teacher Education in Scotland, can function as an effective facilitator of pupils' learning, is committed to professional development and reflection and is able to engage collaboratively with colleagues in the profession, with other groups and agencies, and with various members of the communities served by education
> . . . The benchmark information therefore specifies the standard of skills, abilities, knowledge, understanding and values which programmes should address and assess. (QAA, 2000, p. 1)

The document lists twelve 'core professional interests', including these three:

- promoting equality of opportunity among all people in an inclusive society, and actively taking steps to counter discrimination;
- taking responsibility for and being committed to their own professional development arising from professional reflection on their own and other professional practices;
- *using research and other forms of valid evidence* to inform choice, change and priorities in promoting educational practices and progress (italics added).

But in addition to these core professional interests, programmes will also 'be expected to' provide students with an understanding of 'key educational principles'. Eight such principles are listed. Among these, programmes will be expected to:

- draw on a wide range of intellectual resources, theoretical perspectives and academic disciplines to illuminate understanding of education and the contexts within which it takes place;
- encourage students to engage with fundamental questions concerning the aims and values of education and its relationship to society;

- provide opportunities for students to engage with and draw on educational theory, research, policy and practice;
- promote a range of qualities in students, including intellectual independence and critical engagement with evidence.

There are three aspects of professional development which programmes need to develop.[1] These have a number of sub-sections, as follows:

1. Professional knowledge and understanding
   1.1 Curriculum
   1.2 Education systems and professional responsibilities
   1.3 Principles and perspectives

2. Professional skills and abilities
   2.1 Teaching and learning
   2.2 Classroom organisation and management
   2.3 Pupil assessment
   2.4 Professional reflection and communication

3. Professional values and personal commitment. (QAA, 2000, pp. 8–21)

As we shall see, when we compare these with England, we have here a number of statements that, while attempting to define the nature of teaching in a very explicit way, nevertheless, offer a relatively full version of teacher professionalism, being explicit about values and a research underpinning, for example.

England

Compared with some earlier documents on Initial Teacher Training (ITT) from various Secretaries of State for Education, the 2002 publication *Qualifying to Teach* (Circular 02/02, TTA & DfES, 2002) is concise and attractively presented. The earlier documents, six government Circulars issued from 1984 to 1998 (DES, 1984, 1989; DfE, 1992, 1993; DfEE, 1997, 1998a) had all been written in the dry and legalistic language that might be expected in a Circular, but they were also becoming increasingly lengthy, and more detailed and prescriptive in tone.

On the inside cover of *Qualifying to Teach* it is stated that:

> The document is relevant to anyone involved in initial teacher training, including trainee teachers, qualified teachers and those who employ and support newly qualified teachers. All those directly involved in initial teacher training should have access to it. The document should be used to establish a common framework of expectations and will help to promote the highest professional standards for everyone coming into the teaching profession. (TTA, 2004a)

The foreword to the document is signed by both the Secretary of State for Education and Skills and the Chief Executive of the TTA. The Introduction states:

> Teaching is one of the most influential professions in society. In their day-to-day work, teachers can and do make huge differences to children's lives: directly, through the curriculum they teach, and indirectly, through their behaviour, attitudes, values, relationships with and interest in pupils. Good teachers . . . understand that all their pupils are capable of significant progress and that their potential for learning is unlimited. But teaching involves more than care, mutual respect and well-placed optimism. It demands knowledge and practical skills, the ability to make informed judgements, and to balance pressures and challenges, practice and creativity, interest and effort, as well as an understanding of how children learn and develop. (TTA, 2004a, p. 3)

The language of this introduction is radically different from that which characterised earlier circulars in England. It is indeed imbued with the voice of marketing. It is a very good example of the discourse which is said to have characterised the 'spin' of public sector discourse since the arrival of New Labour in government (Gewirtz *et al.*, 2004). This perhaps reflects the fact that it was recognised by then that it is no longer adequate only to regulate, but that particularly in a context of teacher shortages, it is essential simultaneously to promote teaching as an attractive occupation. However, the preceding Circular, 4/98 (DfEE, 1998a), had no such statement about the role of the teacher in society. Thus, while the words have been carefully chosen to create a positive image of teaching, it is a considerable step forward to be explicit about this (similar to the statement of values which introduces the English National Curriculum 5–16 revision of 1998 (DfEE/QCA, 1999)).

The main body of the document is divided into two sections, one setting out the 'Standards for the award of Qualified Teacher Status', the second setting out 'Requirements for Initial Teacher Training'. These are organised under the following headings:

> The Standards for the award of Qualified Teacher Status
>
> S1 Professional values and practice
> S2 Knowledge and understanding
> S3 Teaching
> > S3.1 Planning, expectations and targets
> > S3.2 Monitoring and assessment
> > S3.3 Teaching and class management
>
> Requirements for Initial Teacher Training
> R1 Trainee entry requirements

R2 Training and assessment
R3 Management of the ITT partnership
R4 Quality assurance

Under S1 there are eight statements, beneath the general heading of 'understanding and upholding the professional code of the General Teaching Council for England'. The first four of these relate to demonstrating respect for pupils and their parents/carers, the other four relate to wider professional responsibilities within the school and the profession, including:

> S1.7 They are able to improve their own teaching, by evaluating it, learning from the effective practice of others and from *evidence*. They are motivated and able to take increasing responsibility for their own professional development (italics added).

S2 deals with subject knowledge, but also with the wider framework of the National Curriculum, such as citizenship, ICT, and the SEN (special educational needs) Code of Practice. The final point is:

> S2.8 They have passed the Qualified Teacher Status skill tests in numeracy, literacy and ICT.

In other words, in addition to successfully completing the programme of study offered by the provider, there are three national tests which all trainees must sit and pass in order to gain QTS.[2]

S3 emphasises teachers having high expectations and being well prepared. It also refers to the need to work with other adults in the classroom. There is also reference to a range of equal opportunities issues, including the need to 'take account of the varying interest, experiences and achievements of boys and girls, and pupils from different cultural and ethnic groups, to help pupils make good progress' (S3.3.6) and to challenge stereotyped views and bullying or harassment (S3.3.14).

The Requirements set out what it is that providers must do in order to ensure that they can continue to offer their programmes. The first and second parts outline trainee entry requirements and some parameters concerning the amount of time to be spent in schools on the various programmes. R3 is interesting in that it requires providers to work in partnership with schools, involving them in planning and delivering the programmes, selecting trainees and assessing them. This is a theme to which we return in Chapter 4.

R4 makes it clear that providers have front line responsibility for their own quality assurance, indeed,

> *All Providers must:* R4.1 ensure that their provision complies with the Secretary of State's current Requirements for initial teacher training.

In other words, it is – somewhat tautologically – a requirement that providers comply with the requirements! There is no reference at this point in the document to how that compliance is actually externally checked through the processes of inspection by OfSTED (although that is mentioned in the introductory sections of the document).

The conciseness of the main document is compensated for by the 107 pages of small print which constitutes the accompanying *Handbook of Guidance*. The uneasy settlement between prescription, control and regulation on the one hand and professional autonomy, innovation and creativity on the other is very evident in the Introduction to the Handbook:

> The Standards and Requirements in *Qualifying to Teach* give providers increased flexibility in the way they design their programmes, and encourage increased use of professional judgement. The purpose of this Handbook is to help providers as they exercise this judgement and to help them maximise the opportunities for development that *Qualifying to Teach* offers. The guidance is non-statutory and will be kept under review. (TTA 2004b, p. 1)

This document is clearly designed to be read by providers and sets out in considerable detail and using exemplars, the fuller implications of each of the Standards and Requirements.

Comparison of documents
In comparing the Scottish and English official definitions of what it is to become a teacher, we can identify a number of similarities, a number of differences, but also some aspects that appear to demonstrate both similarity and difference.

In both countries there is a clear statement of the role of the teacher in schools and in the broader society. There is a broadly common use of language concerning standards and achievement, a shared view that better defined stipulation of what new teachers should be able to do will lead to improved achievement and attainment in schools. Standards sometimes refer to the new teachers themselves but often to pupil performance as well. The documents also share a concern with curriculum, beginning teachers' subject knowledge and assessment. Classroom management skills are a high priority in both sets of documents.

Areas which demonstrate similarity and difference simultaneously include entry requirements (which we have not described in detail above), where it is certainly the case that minimum qualifications in maths and English are set out in both documents, but the Scottish requirements for secondary subject entry are by and large much more precise than those in England.

Finally, some of the most apparent differences include the following. The deployment of 'spin' in the English documents is evident (Gewirtz *et al.*, 2004); the language of marketing is much more apparent in *Qualifying to Teach (TTA, 2004a)* than it is in either of the Scottish documents. The description of the process of becoming a teacher is as 'training' in England whereas it is a process of 'education' in Scotland; this was perhaps inevitable once school-based training was introduced in England and so the learner teachers could not be described as students in any conventional sense of the word. Thus, all learner teachers are described as trainees in England, whereas they are all students in Scotland, every one of them being registered at a university. There is a much more explicit commitment to a strong intellectual component in the process in Scotland than in England, including frequent reference to research and theory in Scotland (the previous English document, Circular 4/98 (DfEE, 1998a), did make reference to research but it has been removed from Circular 02/02 and replaced by a more ambiguous reference to 'evidence' as pointed out in italics earlier). Similarly, there are stronger, explicit statements about teachers' commitment and values (particularly in the areas of inclusion and equality) in Scotland, including statements on teachers' responsibility to combat discrimination; in England the equivalent areas are couched in much milder, less 'politicised' language. There are national 'skills tests' *in addition* to the ITT programme in England, which may be seen as demonstrating a lack of trust in providers' ability to ensure that the programmes themselves can guarantee the development of these skills in all trainees.

There appears to be greater recognition of ITE as the first stage of a professional development continuum in Scotland through the use of interrelated sets of standards at various phases of professional development (especially since the introduction of the one-year induction scheme to replace the former two-year probationary period), with Standards for Full Registration, Chartered Teacher and the proposed Standard for School Leadership (Headship), all being based on similar approaches. The Standards for QTS, Advanced Skills Teachers and Headship in England have not emerged from an overall framework for career development.

Finally, it is worth noting that a recent report commissioned by the European Commission (Eurydice, 2002) on the initial training and transition to work of teachers in Europe (lower secondary) underlined the special importance attached, in Scotland, to the acquisition of communication skills in relation to behaviour management and to cross-cultural communication specifically. Of the few countries in which training in work with multicultural groups of pupils is envisaged, Scotland has the most detailed recommendations with particular emphasis placed on the sociology of migrant populations and methodology of cross-cultural learning as well as the provision of practical training in multicultural classes. Both England and Scotland however, are among the few countries to place particular emphasis on practical training in behaviour management and school discipline as well as general management and organisational skills.

In summary, it appears from this analysis of the key documents that the overall nature of teacher professionalism in the two countries appears to be based on a more 'extended' approach in Scotland with a relatively 'restricted' approach in England, to use Hoyle's early distinction (in 1972, as reported by Stenhouse, 1975).

Discourses of teacher educators

To what extent are these similarities and differences played out in the discourse of some of the key actors in the processes of ITET on either side of the border? In this section we draw on our analysis of the interviews we held with providers of teacher education. The definition of this term could be problematic in that in some sense all partners in the process contribute to the provision. However, here we are referring only to those whose main role is teacher education, whether they be based in a university or in a school.

Scotland

We have already indicated how the general pattern of provision in Scotland is very homogeneous compared with England, with all provision being based in just seven university departments. However, in spite of this homogeneity – or perhaps because of it – we did find, among the three cases that we studied, that each provider does seek to claim some element of distinctiveness.

The pressure on university-based teacher education staff to do research appears to be more prevalent here than in England however and perhaps better supported, at least in some institutions. As one Scottish HE tutor said:

> I mean in terms of development I am undertaking an EdD at the moment and the university is supporting me in that so I am learning quite quickly an awful lot about research in education, plus education theories, all that sort of thing so I am learning a lot.

Another tutor said:

> Yes, it's in my teaching research plan so I've got to do it at some stage. I've probably been more orientated towards doing enough research and writing which will get me through my present situation as probationer and move on to do a doctorate. I mean if I want advancement then I'm going to have to bite the [research] bullet.

And there is a greater consciousness of a model of professional knowledge, drawn from *The Standard for ITE* (the three elements listed above), which while seen as constraining by some, is also seen as a very positive development by a number of teacher educators. As one of them remarked:

> I think both the academic thing has been strengthened and that has been helped by the new professional profile for ITE in Scotland, with a triangle of professional knowledge and understanding and practical professional skills and the values, and the values sit with traditional

study in universities. Because it is a professional course we have these three elements and I think that gives strength to the academic [dimension] and what it is to be integrated, and I think this is a very good way forward.

There is also common reference among Scottish HEI staff to the continuum of professional development:

> I am teaching on the MEd course so it's good because it does flow through and you see the undergraduates doing their initial teacher education and then you can see the teachers continue, which is a life long learning, quite important. I think it is quite an important vision for the [Department] as a whole because we do start with the initial education for school education and we do continue all the way through.

The reflective practice theme is prevalent in Scotland (as we shall see it is in English HEIs) and, in distinction from England often appears to be developed in more concrete ways, including an emphasis on curriculum integration, interaction between theory and practice and a holistic view of teaching. For example, in the words of one tutor:

> There is a strong self-assessment element and it is largely about personal reflection, evaluation, taking things forward, the reflective practitioner if you like and then of course the tutor also gives feedback and that feedback is very much geared to what is being done well and how we are going to take things forward.

Or, in the words of another:

> First part of the lecture programme, the first half of semester is theoretical in nature, we look at intelligence, we look at assessment, we look at differentiation and try to challenge the students theoretically on their experiences the previous semester but also have an eye to their academic progress and what we need to provide for them academically [and] pedagogically. After the mid-semester break there is a shift in emphasis from the theoretical to the practical and a great deal of the second half of that semester is a practical introduction to life in schools.

To summarise what has emerged from the review of our Scottish data, it is certainly the case that there is a confident belief in the importance of academic and intellectual development during the process of becoming a teacher, as well as a strong recognition of the need for practical skills development. The day-to-day experiences are, as in England, of demanding work, with pressures to meet several agendas, but there is also some real conviction that the contribution of the university is a crucial one.

England

As has been indicated already, there is considerable diversity of provision in England, thus part of our interest is to establish whether there is an overriding shared view of the nature of teaching and of what counts as professional knowledge or whether, on the other hand, each form of training tends to relate to a distinctive discourse of professionalism.

One of the team members of the SCITT visited used language which is not only more individualised than that used in universities, it is also imbued with teaching as 'folk knowledge':

> The thing is with classroom management, it's all to do with relationships, isn't it, I mean there will be the odd kid here or there, that, when I say lost causes I don't mean that, but you know whatever you do you're not going to win. But on the whole, if you've created the right relationship, and you actually tailor your curriculum to the needs of the children, you don't have many behaviour problems. And a lot of teachers who are practising every day in school, haven't realised that.

Or in the words of the manager of the SCITT:

> We talk about the key features, we talk about progression, we talk about teaching methods, which I am sure are all bog standard, we then tend to put them into schools fairly early, different from the HE experience here anyway, where effectively they're doing the theory during the first term, and then they go into the practice.

He was very conscious of the need to keep people on track once they had started the programme:

> We now bring them back [from school] so that we can check on how things are going, and react quickly if it isn't going well. Because these are very valuable commodities, and we don't want to lose them if we can avoid it.

The manager of the Employment-Based Routes (EBR) consortium makes reference mostly to class management and also to subject knowledge, but said nothing about reflectivity or theoretical underpinning. For him there is little doubt that recruitment was driving the agenda (he came from an LEA background) and the training was organised on a pragmatic basis, aiming simply to ensure that people who were already working in schools were able to achieve a formal teaching qualification. Indeed the programme appears to be very much about whether the trainee 'fits in' to the school setting. Talking about the initial judgement as to whether to accept someone onto the programme, he refers to the period of time a trainee spends under observation in a school:

It becomes fairly evident fairly quickly whether that person is going to be an asset or a liability and if that happens, if that person has made that commitment, if they have come through, if the school is saying welcome aboard and now we are trying to get you a place on the programme, that's how I want it to work, and therefore it's different from ITT.

However, his colleague, who was a longstanding university lecturer who was providing some of the training input to the scheme, had a rather different view. He indicated that he had found a new kind of freedom working on these programmes, one which was not constrained by the curriculum of conventional ITT courses, but could be influenced by what the trainees themselves brought to the programme:

> What I try to do with both the graduate teacher groups and the registered teachers is to use their knowledge so it's not me, I'm just kind of directing in that sort of sense and add in what I feel is appropriate, so they will contribute and they tend to be in a very positive way of, yeah that's reinforcing what they've done already.

The priorities that the graduate trainees bring to the programme are extremely pragmatic; they are not seeking further intellectual development it would seem. He said:

> I deliberately approach the groups differently because of different sort of follow up formally or informally, so in a sense with a graduate teacher, well it's common to every programme but all they would want if they wanted more sessions would be more classroom management. That's it, they just want the whole lot on classroom management.

In contrast to this approach, we found that HEI-based staff tended to emphasise links between theory and practice and to refer to particular models of teaching, usually some form of 'reflective practice'. Indeed each of these providers referred to particular conceptual frameworks upon which their programmes were based. These were usually derived from a mixture of experience and research and, while they took the Standards into account, they sought to go 'beyond' them. So it was that many of the university providers made claims that their own courses were distinctive:

> The other distinctiveness is what we argue as value added, because we believe that the standards aren't merely the standards and although they are quite stringent in some respects, we like to think that someone emerging from this university has got the standards plus something else. The something else we would normally see as being to do with reflective practice and ability to analyse and understand

where they are at, have a sense of how educational research can play a part in a typical classroom teacher's delivery. We are not interested in divorcing the classroom practice from research or enquiry and we see that as being instrumental in altering and affecting the quality of pedagogy that goes on in schools, so in a sense we reach out to the CPD agenda whilst we have got trainees on ITT programmes.

He also argued that the university contribution is crucial:

We have to keep bearing in mind that we are a University provider and we are providing an award here, we are not just providing the standards. If somebody wants to just do the standards then they can go and do the GTP, if they want a PGCE then it has to have some kind of additional merit to it hence the assignments that are involved in the PGCE. Yes, they cohere with the standards and they partly deliver the standards, but it also forces the trainee into far more intellectual and academic understanding of the job of a teacher, where they develop this sense of professional understanding, they have a critical eye. I am not convinced that kind of criticality comes through a GTP course where those kind of academic assignments simply aren't required.

There is also evidence of continuing commitment to child-centredness, the promulgation of which had been a target of the strident attacks on traditional teacher education in the 1980s, referred to above (see Chapter 2). As one such teacher educator commented:

So I suppose the central plank would be the child at the heart of it all, not the curriculum but the child. So I think if you pressed our students I hope they would think that the child was absolutely central to what was going on here and that is their first priority. They would try, I think, to be as specific in terms of differentiating the work to make this appropriate for the child, they would think about the children I hope as individuals rather than just a class.

Much of the discourse from all English providers was imbued with an awareness of the 'culture of compliance' which had been created during the 1990s by the twin operations of OfSTED and the TTA. There was frequent reference to Standards, as in the case of this college lecturer:

We work within particular constraints. Clearly there are certain things even under *Qualifying to Teach* (02/02) that have to be in the course in order for us to reach the standards for both courses at the end of the time allocated to it. Those are a given in a sense.

There is also a sense of competition and vulnerability within the system, as well as a desire to keep the constraints 'under control':

Universities and colleges are not going to come together, regionally and nationally, because of the OfSTED agenda. Each institution is responsible for its future and we're not going to change our procedures if we think that's going to be detrimental to our OfSTED outcome. You can see the dilemma, can't you?

So in summary, there does appear to be considerable diversity in how professional knowledge is constructed on different programmes in England. Those responsible for the HEI-based programmes demonstrate a commitment to the integration of theory and practice and acknowledge the significance of the school-based element. The standards tend to be seen as a series of hurdles to be got over, rather than the basis for a coherent or complete model of what it is to be a teacher. The university providers aim to add value beyond the standards. On the basis of the evidence in this study, it appears that the approaches which are not so influenced by HEI staff, such as SCITTs and EBRs have much less aspiration to promote theory, research and integration in their model of teaching. They are in a sense both less ambitious and perhaps more honest.

Comparison of teacher educators' discourses
There are many similarities in the discourse of teacher educators in Scotland and those teacher educators in England who are based in universities. The university staff in both England and Scotland share concerns about the relationship between theory and practice. They also are very aware of the regulatory framework of standards or competences which are required, but whereas the English frequently refer to the next or the last visit from OfSTED, there appears to be greater self-confidence among the Scottish providers.

Indeed, in terms of the understanding of what teacher education is trying to achieve there appears to be a greater similarity between the university-based staff on both sides of the border than there is between the English school-based provision and the English universities. Our coding system for data analysis highlighted two strands in each interview – strand one concerned organisational aspects of the programmes such as course structure and partnership arrangements, whereas the second strand was about curriculum and professional knowledge. It was very noticeable that in England the second strand was far less evident in the interviews with SCITT or EBR leaders than in those with HEI tutors.

While in England most providers expressed a strong sense of anxiety relating to the inspection regime, a similar level of anxiety was expressed by some teacher educators in Scotland about achieving sufficient personal research development.

The discourse of reflective practice was found on both sides of the border, although it appeared to be more fully developed and articulated in Scotland and less frequently deployed in school-based approaches in England. However, as we anticipate the review of partnership matters in Chapter 4, the relative underdevelopment of the role of schools and teachers in the process of ITE in Scotland appears to be something of an impediment to the full integration of professional knowledge across the ITE experience for students there.

Conclusion

If our analysis of the key policy documents and teacher educators' discourses in ITET in Scotland and England is correct, then how might the differences in the ways in which teaching and teachers' professional knowledge are defined best be explained? Much scepticism has been directed towards the 'mythology' of Scottish education which makes the dual claim that Scottish education is superior and that it is more genuinely 'comprehensive' (i.e. open to all) and/or meritocratic (see, for example, Humes and Bryce, 2003; Paterson, 2003). Yet, it is difficult to offer an explanation of the more extended professionalism and the greater academic component in the way in which teaching is seen in Scotland, without reference to the relative cultural standing of teaching, teacher education and education in the two countries (we return to this theme in Chapter 5).

However, we should question whether the adoption of a standards/competences approach in both countries is of greater significance than the differences which we have identified. Certainly the notion that an intellectual and creative process such as teaching can be described, indeed defined, through a list of observable behaviours does imply a form of 'technicism'. Indeed criticism of such 'technicism' has been put forward forcibly in Scotland as well as in England (see Stronach *et al.*, 1996; Furlong, 1995 or Whitty and Willmott, 1995). The continuing commitments to theory, research and social justice which appear to characterise the Scottish approach may in reality be but a modernist palimpsest on an underlying instrumentalism, whereas, at least in its most recent manifestation, something equivalent is provided by the careful branding ('glossification') of teaching in England. It is certainly the case in both countries that there is little encouragement to take risks or to be experimental in one's approach to teaching, whether at the beginning of one's career or later.

What emerges in that analysis is that the discourses employed by university-based educators on either side of the border consistently show more explicit commitment to values such as those described above (often articulated through a notion of 'reflective practice' in teaching) than the discourse deployed by school-based teacher trainers in England (i.e. those working on SCITTS and EBRs).

When we compare the two initial teacher education/training systems, the nature of the involvement of the university sector does appear to be a defining factor. In England the OfSTED reports on school-based approaches have indicated some concerns about the quality of provision (e.g. OfSTED, 2002). In principle the creation of standards and competences is meant to ensure consistency of achievement for all teachers, but do those who have engaged with an HE-based approach during their preparation demonstrate any additional features and/or are the competences that they develop applied in a different way?

In an earlier consideration of English and Scottish approaches to the 'modernisation' of teaching, in the context of *established* teachers who are several years into their career, the evidence suggested that the Scottish approach was strongly 'developmental' by contrast with the 'performative' and 'assessment oriented' approach south of the border (see Menter *et al.*, 2004). In considering now the way in which teachers are *initially* formed on either side of the border, we can see some similar tendencies, but also some interesting differences from that earlier comparison. Perhaps because this pre-service stage of teaching is necessarily process-oriented we can see much more recognition of a developmental approach in England than was evident in looking at policies relating to experienced teachers. But at the same time we can also see a much more performative element in the Scottish arrangements than was the case when looking at later career development. This may be explained by the establishment of common systems of standards in both countries that require some form of outcome criteria, for the purpose of ensuring that individuals have achieved the required level. Convergence of this sort could be seen as a response to a perceived need for more uniform patterns of training that would correspond to national and international quality standards as well as facilitate quality control itself (Eurydice, 2002, p. 41). Indeed this convergence is very much in line with the theme of *accountability* that was foreshadowed as part of the globalising tendencies identified in Chapter 1.

However, even though the processes are broadly similar, the intended outcomes, both as expressed in official documents and as expressed in the words of teacher educators do appear to display some significant differences. While technical rationality has had its impact in both countries, a traditional modernist discourse continues to be visible in Scotland, whereas in England, technical rationality is only slightly obscured by traces of post-modern packaging. In neither country do we see the full blown radical reconstruction of teaching along the lines which Parker (1997) has argued is necessary to meet the demands of the post-modern world in which he at least lives:

> Postmodernism issues in a plurality of educational dialogues, practices, ends and values. This will involve education and teacher-education institutions in becoming less like departments of science

and more like departments of literature; less like the factory production line . . . and more like a fashion house, where the multiplicity of styles coexist . . . (p. 149)

The university remains the most important site of struggle in teacher education. It is clear from the recent history of ITT in England that politicians of a neoliberal persuasion (initially in the Conservative Party, later in New Labour) were determined to reduce the influence of the academy on teaching. The apparently paradoxical move in the opposite direction in Scotland – that is – the amalgamation of colleges of education with universities – is better seen as a rationalising confirmation of the HE base for ITE, which has brought its own problems for those universities (see Brisard *et al.*, 2006).

Young (1998) makes a call for the 'reflexive modernisation' of teaching (not dissimilar to Edwards and colleagues' (2002) call for 'rethinking teacher education'), in which universities would play a crucial role:

> The responsibility of those based in universities is not just to critique the bureaucratic character of recent reforms. It is also to articulate real alternatives and how they can raise standards and support a new teacher professionalism which puts learning at the centre of the curriculum of teacher education. (Young, 1998, p. 167)

On neither side of Hadrian's Wall[3] can we see learning being put at the centre of the teacher education curriculum, in spite of the frequent references to 'the learning society' and the 'knowledge-based economy' by politicians and others. Such concepts, Ball argues, are but 'potent policy condensates' which reflect the increasing colonisation of education policy by economic policy imperatives (Ball, 1998, p. 122). What we do see is a common urgency to be precise about what it is that teachers are required to know and to be able to do. There is a drive in both countries towards 'standardisation'. The significant differences between the two sets of definitions of the teacher reflect a different nexus of power and control in each country and a different cultural positioning of teachers, teaching and, indeed, public education.

Notes

1. These three aspects are also used in the definition of other standards, including those for Full Registration (at the end of the induction period), Chartered Teacher and School Leadership.
2. See Mahony *et al.* (2001) for an appraisal of the significance and impact of these tests.
3. Hadrian's Wall is an historic (Roman) barrier between England and Scotland.

Chapter 4

FROM PRACTICE TO POLICY?
THE PROBLEM OF PARTNERSHIP

Introduction

As we have explored elsewhere, the development of models of partnership between higher education providers and schools within ITET programmes has been a major international area of activity in recent years.[1] The current analysis also draws upon work which we have presented elsewhere, specifically considering the recent history of partnership in Scotland (Smith *et al.*, 2006a) and the comparative development of models of partnership between Scotland, England, Northern Ireland and Wales (Smith *et al.*, 2006b) as well as from our Scottish 'Convergence or Divergence' findings relating to partnership (Brisard *et al.*, 2006). Beyond this, the present chapter broadens our use of 'Convergence or Divergence' data from Scotland, and considers fresh data from England, in supporting the analysis of partnership developments.

Models of partnership

The typologies of ITET partnership produced by writers such as Furlong and colleagues (2000) can be taken as useful frameworks of reference for analysing partnership developments in Scotland and England. Furlong *et al.* have identified four models of partnership.

The first of these, which would have been found in England and Scotland in the early 1990s, can be described as the *HEI-based* or *integration* model (sometimes referred to in Scotland as the *duplication* model, see Cameron-Jones and O'Hara, 1993, p. 37). In this model, the roles and responsibilities assumed by HE tutors overlapped with those assumed by teachers and partner schools, with HE staff attempting to 'integrate the students' training experience in college or university with the world of the school' (Furlong *et al.*, 2000, p. 76). This meant that HE tutors presented campus-based sessions which were highly oriented to practical needs, explicitly modelling school classroom teaching, and they set assignments which were based on the work the students undertook in school. The HE tutors took overall responsibility for course planning and assessment, including taking responsibility for assessment of the students' practical teaching competence on school placements. This was through HE tutor assessment visits to schools, with school staff themselves assuming minimal formal responsibilities for student assessment. Aspects such as these HE tutor roles were highlighted by Cameron-Jones and O'Hara as examples of the duplication between HE tutors and schoolteachers involved in such a model of partnership.

Furlong et al.'s second model was the *complementary* model, sometimes also described as the *separatist* model, which he saw as developing from the early 1990s in England. The key distinction between the complementary model and the previous HEI-based model was that the complementary model attempted to establish a clear separation of distinctive roles and responsibilities for HEI staff and school staff. For example, the duplication of the HEI-based model in relation to the assessment of the student teachers' classroom practice was removed through school staff assuming formal responsibility for this assessment, with HE staff reducing their school assessment visits to at most 'troubleshooting' or moderating visits. Similarly, in the complementary model, HE staff would seek to make their campus inputs more explicitly theoretical and research-based, rather than duplicating the more practical preparation sessions which could be covered by school staff with student teachers. The separatist nature of this model lay in the requirement upon the student personally to integrate their essentially separate higher education work and school-based work.

Furlong et al.'s third model of partnership can be described as the *HEI-led* model, which they saw as developing in England during the mid to late 1990s. For these authors, the HEI-led model differs from the complementary model because the HEI renews a sustained effort to provide overall leadership for both the HEI-delivered and school-delivered elements of courses, e.g. taking clear responsibility for overall planning of all elements of courses. The HEI-led model can also be clearly distinguished from the earlier HEI-based integration models because school staff will now have agreed formally to accept specified roles and obligations within an HEI-led partnership, for example, in relation to the assessment of student performance during placement.

For Furlong et al. (2000), there is a fourth model of partnership: the *collaborative* model. In the collaborative model of partnership, a dialectical approach to theory and practice is encouraged by stressing reflective practice in the student teacher, with the student teacher drawing upon the different forms of professional knowledge contributed by staff in higher education and staff in schools, but with both of these seen as equally legitimate. This model requires regular opportunities for HE staff to meet with school staff for small group planning of programmes and for collaborative work and discussion during HE staff visits to schools. Writers who favour the collaborative model give particular stress to its central emphasis on an equal relationship between higher education and schools, for example, as described by Sachs (2003, p. 66) 'a two-way model of reciprocity [which] assumes that each party has something to contribute to the professional learning of the other'. However, as will be suggested, the collaborative model is generally seen as aspirational within the UK, rather than widely achieved.

In a sense, the aspirational status of the collaborative model within the United Kingdom can be seen as a point of convergence over partnership between Scotland and England. Certainly, there have been individual examples within England of successful collaborative models of partnership. The most prominent of these has probably been the Oxford Internship Scheme for the Postgraduate Certificate in Education (PGCE) for secondary teaching (see McIntyre, 1997). However, such fully-collaborative models have not been widely adopted in England, and no fully-collaborative models of partnership have been developed at all in Scotland. Despite apparent convergence on this point, the narrative of partnership development, and the current position on partnership within Scotland and England can be suggested to diverge in a number of significant ways.

Partnership in Scotland: an overview

We have detailed elsewhere our overview of the development of partnership within Scottish ITE from the early 1990s (see Smith *et al.*, 2006a; 2006b). The narrative which we would present is of a Scottish higher education community within ITE which recognised the limitations in the early 1990s of the HEI-based or duplication models of partnership, and was favourably disposed to the development of complementary models of partnership, ultimately leading to collaborative models (for examples of these views, see Elder and Kwiatkowski, 1993; Cameron-Jones and O'Hara, 1993; 1994a; 1994b; Cameron-Jones, 1995). The Scottish Office Education Department (SOED) then launched the Mentor Teacher Initiative, a clear attempt to develop complementary partnership approaches by identifying a formal, enhanced role for school staff in supporting and assessing student teachers. The Initiative was piloted at Moray House Institute of Education in 1992–3 and 1993–4 with the Institute's PGCE Secondary Course. The Mentor Teacher Initiative, if successful, would have begun to move Scottish ITE onto complementary models of partnership. However, to the considerable frustration of many within the HE sector, the Mentor Teacher Initiative failed. Essentially, the Initiative was abandoned by Scottish Ministers in October 1995 in the face of resistance from Scottish schoolteachers (see Kirk, 2000, p. 43). As their strategy for resolving the political impasse over the Mentor Teacher Initiative, Scottish Ministers asked the GTCS to set up a Working Group on Partnership in Initial Teacher Education. The Working Group first met in December 1995 and its Report was published in March 1997 (GTCS, 1997). The Report of the GTCS Working Group on Partnership did not succeed in unblocking the policy impasse created by the failure of the Mentor Teacher Initiative. While the GTCS Report did identify involvement in ITE as a general responsibility of the teaching profession, it linked this involvement to the insistence that additional resources would be required in schools, if school staff were to develop fuller roles within ITE partnership.

With issues of partnership development unresolved by the publication of the GTCS Report, further developments of partnership models in Scotland became dependent upon how the public policy context itself progressed. Following the publication of the GTCS Report, the Scottish Executive engaged in a number of internal exercises over partnership in ITE, some of which produced documents in the public domain. For example, there was a review of the cost of partnership undertaken by Deloitte & Touche, jointly commissioned by the Scottish Executive and GTCS (Deloitte & Touche, 1999). Following the McCrone Committee of Inquiry into Teachers Pay and Conditions, and the subsequent McCrone Settlement (Scottish Executive, 2000; 2001a), the SEED commissioned Deloitte & Touche to conduct a First Stage Review of Initial Teacher Education, with the consequent report being produced in June 2001 (Deloitte & Touche, 2001). Despite an Action Plan being released following the publication of the First Stage Review of Initial Teacher Education (Scottish Executive, 2001b), no conclusive outcomes were progressed before the Scottish Executive announced a Second Stage Review of Initial Teacher Education in 2003 (Scottish Executive, 2003). This led to a Report which was finally published in 2005, together with a Ministerial Response (Scottish Executive, 2005a; b). Certainly, the Second Stage Review of Initial Teacher Education includes the positive stress on the need for 'new effective and pro-active partnership' between local authorities and universities, with 'local authorities being more actively engaged in ITE'. This theme is also emphasised in the Ministerial Response which accompanied the publication of the Second Stage Review Report. However, the Second Stage Review Report, and the associated Ministerial Response, tend to concentrate upon the operational organisation of student placements within partnership, e.g. with the emphasis on local authorities taking a more strategic co-ordinating role in identifying student placement opportunities to maximise the capacity for student placements. There is much less discussion in these documents of the underlying roles and responsibilities of the wider schoolteaching profession within ITE, especially on the respective roles and responsibilities of school staff and HEI staff in partnership. In this sense the Second Stage Review can be suggested to leave unresolved fundamental issues which have been a source of tension in any attempt to develop models of partnership within ITE in Scotland since the early 1990s (see Smith *et al.*, 2006a, for a fuller development of this analysis).

Partnership in Scotland: the evidence from stakeholder discourses

Consistent with the general overview, the evidence from stakeholder interviews conducted as part of our 'Convergence or Divergence' research project illustrates the unresolved nature of partnership development in Scotland. We have analysed much of this evidence elsewhere (Brisard *et al.*, 2006), but we would again highlight key stakeholder perceptions of the current state of ITE partnership in Scotland.

As was the case with their predecessors in the early/mid 1990s, Higher Education providers of ITE in Scotland can still be shown expressing strongly their desire to clarify the roles and responsibilities of school partners as part of a move towards a complementary model of partnership.

One HEI member of staff commented:

> We really need to get on some sort of sound basis for the notion of placements and where the schools feel their professional obligations, professional commitments want to be in relation to that. That is just causing chaos at the moment, there is no doubt about that [we need] a genuine shared understanding of what we are meant to be doing on a complementary basis.

HEI staff can also be found advocating strongly their aspiration to move toward the type of relationship with school staff central to the best models of collaborative partnership.

As another HEI member of staff commented:

> Instead of tutors going in and watching students teach, and assessing them, you can have tutors going in working alongside students; you can have tutors going in working with the teachers doing development there in terms of how to mentor, how to support; you can start looking across boundaries so that instead of seeing ITE in isolation, you see a developing role for the mentor, not only across ITE but also across the Induction year and into the Standard for Full Registration in terms of continued professional development.

On the other hand, in contrast to these aspirations, HEI staff are to be found expressing their frustration at the current partnership situation in Scotland. One Head of a Faculty/School of Education in Scotland emphasised his dissatisfaction with the failure of other stakeholders, specifically teachers in schools and local authorities, to assume the kind of roles and responsibilities necessary to develop complementary and collaborative forms of partnership. Regarding the teaching profession, this Head of Faculty/School commented:

> If you say there's no reason at all why HEI staff should have to engage with some of the practical preparation and the short-term evaluation

of classroom teaching, you're going to get a negative reaction from the elected teachers of the GTC, partly because these people are also trade union activists, so their perception about the workload implications on that will be very central. It's not their [teachers'] responsibility to assess the student, it's the HEI's responsibility.

Regarding local education authorities, this Head of Faculty/School commented:

[what the education authorities say is . . .] what we really want is the research [from the universities]. Then if you say to them, well you can only get quality research in Scotland if you guys do something to help us [and] get your staff to take over responsibility for supervising placements so that my staff are free to get involved with research, they'll not help you, we have to do a great deal of work on that.

On the other hand, teachers can be seen expressing a commitment in principle to assuming fuller roles and responsibilities within ITE. However, this is never unconditional and will always be linked quickly to resource issues. For example, one senior headteacher, significantly involved in ITE policy and practice at national as well as school level, commented very positively upon the teaching profession's general interest in ITE, but immediately linked this to the profession's concern about resources:

I would say that by and large the teaching profession is very positive about ITE. If there is a frustration at all, it is about having the resources to actually work closely with student teachers. There is a spirit of co-operation in terms of student placements in school. There is an anxiety in the common resource to do that.

This headteacher went on to emphasise that the partners in ITE

at the moment just don't have the time to do it and you will find that willingness to be involved probably will be directly related to the resource that people have . . .

He contrasted this with the resourced support which probationary teachers in Scotland now received under the one-year induction scheme:[2]

I think that the whole issue of student teachers, particularly placements, should have been wound up in some sort of package, to have a very similar deal to what we have with the probation of staff just now. The fact that probationers are now well supported in schools, by professionals in schools, now makes that difference much more stark.

Of course, even this conditional support for a major role for school staff within partnership came from a senior headteacher significantly involved in

ITE policy and practice, both nationally and locally. Certainly, in our wider interviews with a range of school staff directly involved in working with students on placement, schoolteachers are not generally rejecting involvement with ITE students. There is a recognition that teachers have some kind of professional duty to be involved, linked to the extrinsic expectations of other stakeholders that this is part of their role. One teacher explained that his colleagues worked with students 'Because they are professionals. They know that we have to do this.' Another commented that 'there is a commitment that we are required to give this'.

However, we do find that the language used to express this professional duty can intimate some reluctance and an enthusiasm which is qualified. As one teacher somewhat conditionally conceded about his involvement with ITE students 'from a professional point of view I do feel that they need to come somewhere'. Scottish schoolteachers generally are also quick to stress what they see as current additional pressures on their capacity to work with student teachers.[3] Of course, this somewhat qualified commitment is only to the present roles and responsibilities assumed within partnership.

Our wider interviews with school staff showed little evidence of engagement with deeper issues of partnership development. Perhaps understandably, even staff very experienced in supporting students showed little interest in the broader aspects of partnership. For example, one experienced secondary head of department, when asked about the general course documentation which the partner HEI had sent before placements, commented: 'I probably read it and not found it as particularly important to my job, and therefore it had to be brought to my attention.'

Another experienced head of department from a different partnership stressed that he drew upon his own experience rather than the HEI documentation, when working with student teachers:

> Really I just draw upon my experience. The guidelines, and obviously the forms we have to fill in at the end of a placement, direct us to specific areas which obviously we are concentrating on as they are going through their time here. However, mostly I draw from just my own experience in the classroom and my experiences of observing other people teach as well.

These views did reflect wider comments where teachers emphasised a very practical approach to the ITE process, without necessarily criticising the HEI role. As one secondary teacher responsible for ITE students in his school commented:

> I certainly feel that the more time that can be spent in school on-job-training the better, although I'm not in any way trying to demean what the HEIs do.

On occasion, we found someone who went beyond a lack of engagement with the HEI's approach to partnership and displayed a more confrontational opposition to the HEI's 'theoretical' contribution. For example, one senior member of staff with overall responsibility for students in his secondary school commented on his priorities for the ITE experiences of student teachers:

> Things like, how well do you know how to teach a subject, can you communicate with children, can you manage a classroom? These are things which we regard as slightly more important than that you have read the latest research on education which may have value in its way, but when they are just trying to get started in this job you need the craft side of teaching rather than the theoretical side of teaching.

This teacher went on to suggest that his partner HEI was guilty of 'the elevation of theory above practice which I think is quite dangerous for what is a practical job at the end of the day'.

The unresolved nature of partnership within Scotland can also be evidenced in the views of ITE students. Certainly, students can be found supporting the aspirations of a complementary approach to partnership. One student drew upon the views of her tutor to support the distinctive contributions of both HEI and school staff:

> My subject tutor said to me you will learn most on your practices, the university is there to try and support you, to try and guide your thinking . . . you are going to learn to teach in a school, which makes sense, and you are going to learn about the theories and reasons behind why they do it in the university, and I do think that it is important that you learn about the theories behind it.

Students also recognise the contributions which university staff can make to partnership and the student's experience within it:

> I know my tutor has been off teaching for a long time, but he's done lots of research, he was really helpful and he gives us lots of new things to teach, new strategies. I was on placement and I said 'Can I do that?' and the teacher was 'Oh really, how exciting, good' so I think both parts were really helpful for me.

Another PGCE student summarised:

> Yeah I think it is very important to have the 18 weeks in college, because we're going on placement for the rest of our lives, so I think it is important to have those 18 weeks. I think very much they (HEI staff) instil that we have to be reflective and think how you can improve it yourself. And I think if you take that as a basis for your training and carry that on into your career, they are certainly giving us the basis to build on that.

On the other hand, students also recognised the inappropriate duplication which could be involved in current roles and responsibilities. For example, one student graphically highlighted the limitations in the current model of tutor assessment visit, as compared to the on-going assessment by partner school staff:

> The tutor really doesn't know that a certain child who is maybe not doing anything, or is working very slowly but is working, that is actually ok for them, whereas normally they might be lying sprawled on the floor or something. They only got an hour or a wee bit more to really judge what is going on in the room which is not always that much time, so I think in a way the teacher can actually give you a lot more support.

The current, more limited, commitment of schoolteachers within partnership is also highlighted by students:

> I found that our tutors from university were more interested in us as an individual. They were giving us advice as to how to develop you know holistically, our personality, the whole teacher. Whereas in the classroom I felt my class teacher was just 'how to teach this class'.

Finally, Scottish students can also highlight tensions between the HEI and school approaches within current partnership, again illustrating the limitations in these arrangements:

> I know I find the theoretical side very interesting and I think it was good that we got a grounding before we came into the school, but I think the problem lies where some of the staff in the university are expecting us to do things that just aren't practical in the schools and the teachers in the schools say, 'No that won't work.'

Partnership in England: an overview

As has already been suggested, Furlong et al.'s typology of models of partnership (2000) can in a sense be presented as a historical summary of developments over partnership in England since the early 1990s. Our current overview also draws upon our earlier analysis of the development of English models of partnership in the comparative context of the rest of the United Kingdom (Smith et al., 2006b).

The 1990s in England began with HEI-based or integrated partnerships. England then moved onto developing complementary or separatist models of partnership. From 1992, policies of central government increased the role of schools in partnerships and stressed the importance of clearly defining the respective roles and responsibilities of partner schools and HEI staff (DfE

1992; 1993). This approach was associated with a resource passing from HEIs to schools to recognise the significant contribution of schools to the teacher training process. This approach was initially applied to secondary schools in 1992, but similar approaches were applied to primary schools in 1993. Perhaps more significantly, the 1994 Education Act disbanded CATE, established the TTA and allowed for the direct funding of SCITTs, which may or may not involve the contribution of HEIs.

This strengthening of the role of schools was further extended by the Labour government elected in 1997. The 1998 Green Paper on teachers stated that:

> We want to encourage good and innovative practice in school-led initial teacher training . . . we are determined that the employment-based routes into teaching should be recognised as providing higher quality preparation for entry into the profession. (DfEE, 1998b, para. 111 and 115)

The government Circular of 2002 governing ITT (TTA and DfES, 2002), with its associated Guidance Notes, emphasised that forms of partnership can include not only 'schools working in partnership with HEIs' but also 'several schools working together with or without the involvement of an HEI to provide School Centred ITT (SCITT)' and 'schools working with an LEA, HEI or another school to provide an employment-based route to QTS' (TTA and DfES, 2002 – Guidance Notes, p. 84).

However, while these policy initiatives in England may be seen as developing complementary or separatist models of partnership, they are certainly not seen by stakeholders as moving towards fully collaborative models of partnership. For example, there is considerable concern that these developments have been the result of top-down policy initiatives through the TTA, aimed at increasing central control over the context and arrangements of ITT courses, rather than shared developments taken forward by the stakeholders within partnerships themselves (see Furlong *et al.*, 2000, p. 164). These concerns have also been linked to opposition to the OfSTED inspection regime for ITT, which has been perceived to be burdensome, intrusive and threatening, especially because of its association with decisions about the continuation and funding of specific ITT provision (see subsequent evidence from stakeholder discourses).

There has also been specific concern in England that these partnership developments have tended to destabilise and marginalise the roles of HEIs within ITT. For example, Stephenson (1994, p. 19) highlighted the following key question from Wilkin's early work: 'to what extent is mentoring a reversion to an earlier form of apprenticeship training?' (Wilkin 1992, p. 26). Furlong *et al.* (2000, p. 165) have also commented that partnership reforms, in spite of considerable resistance from the HEIs, have significantly narrowed and weakened the contribution of HEIs to the training process, and indeed in many institutions training has developed an increasingly practical orientation with the relationship between those in schools and higher education narrowed to bureaucratic rather than collaborative relationships. More widely, concern over any such weakening of the HE role can be found in Wilkin's commissioned paper for UCET (Wilkin, 1999) and Furlong's commissioned paper for Universities UK and SCOP (Furlong, 2000).

Partnership developments in England can generally also be seen throughout the 1990s to have involved significant marginalising of any role for LEAs. Certainly the TTA's National Partnership Project of 2001 (see Furlong *et al.*, 2006) was intended to include LEAs in its work with ITT providers in schools through the strengthening of the quantity and quality of ITT partnerships. On the other hand, this approach to establishing nine regional partnerships can again be argued to be a form of 'top-down' co-ordination being imposed centrally on individual local authorities. In any case, overall evaluation of the National Partnership Project has suggested inconclusive progress (Furlong *et al.*, 2006).

Of course, Furlong *et al.* (2000) have also suggested that the mid to late-1990s in England saw HEI-led models of partnership, as HEIs made a sustained effort to provide overall leadership for both the HEI-delivered and school-delivered elements of courses in response to the separatist tendencies of the complementary model. This would see the HEI taking clear responsibility for overall planning and defining of approaches to school placement learning and assessment. Certainly, this can be seen as the HEIs attempting to assert their role within ITT and is in line with the argument put forward by Furlong (2000) on behalf of Universities UK and SCOP. The author argues that it is important to preserve the enhancing role of HE through the involvement of active researchers, and the contribution which they can make to existing knowledge about the process of becoming a teacher. However, the HEI-led approach is still well short of collaborative aspirations. This is because the HEI is exercising its leadership through much more limited ongoing contacts with school staff, for example, concentrated on a small strategy group of school staff, rather than through the wider joint planning and collaborative working which would be the characteristics of a truly collaborative model.

Partnership in England: the evidence from stakeholder discourses

As with Scotland, our interviews with stakeholders as part of our Convergence or Divergence research project illustrate the development of the main themes within partnership in England.

Heads of Faculties/School of Education can be found supporting the aspiration to collaborative models of partnership. As one commented:

> I think the Oxford model that recognised the complementary nature of school contributions and HEIs contributions to teacher education, ought to have more influence than it has in practice. There is not a lot of point in having teacher educators in higher education institutions if they are identical, in terms of their skills, to teacher educators in schools. They might as well be in schools. It is only if we've got something specific that is complementary and adds value that you actually need us. And the Oxford model recognises that.

Interestingly, we also found clear examples of school staff who appeared committed to the creative possibilities of genuinely collaborative partnership, within which school staff interacted positively to the ideas HE staff presented within a 'learning community' context. One headteacher described commitment to ITT in their school:

> It is driven here by a view of education. We see the school as a learning organisation, it is that kind of rhetoric. If we can model internally and create situations where people can legitimately challenge each other's practice and be aware of leading edge practice then that will be powerful for the school.

Heads of Faculties/Schools of Education clearly identify the distinctive contribution of HEIs within complementary models of partnership. As one emphasised:

> When the students are in university, yes that's the right place for higher order thinking, you know that you do study education thoroughly so that you are an educator rather than simply being a trained teacher. You learn to survive in school and that is important, but you couldn't survive in school unless you were equipped with what you get in university and it is not just tips for teachers it is about thinking and reading and being informed about research.

However, Heads of Faculties/Schools of Education also explicitly recognise the move to fuller school responsibilities within complementary models of partnership. One commented:

> I think that there is an expectation on schools and schools know what they're expected and required to do and I think that is good. I think it is good that students should know what their entitlement is and should be able to ask for that if they are not getting it and should be supported in getting it. So I think there are a lot of things which have changed for the better as a result of the accountability and the structures that have come about.

Another respondent highlighted the willingness of teachers to assume enhanced responsibilities, noting that 'there has been a transformation in the interest teachers show in training students, they want to do it and some regard this now as their career path'.

HE tutors also comment on the increased role of schools within partnership. As one remarked:

> I think also the balance between the part that's done by the HEI and the role played by the schools has also altered, and I think we've moved very much towards the role of the schools.

School staff involved in partnership with HE institutions also indicate a willingness to assume responsibilities within partnership which previously may have rested with HE staff. As one teacher responsible for placements within their school commented on the local partnership with HE:

> The student's individual HE subject mentor didn't come at all. Who came was the link tutor. The link tutor was responsible for making sure that things were going OK and going into school occasionally just to see that things were alright. Then the school provided the report in the end at that time.

Even school staff more normally involved in GTPs/SCITTs comment on the clear responsibilities which they expect to assume if accepting student teachers on an HE programme. One such teacher emphasised their school's formal commitment in these cases:

> The point I'm making is we had to sign a partnership agreement, with all the ins and outs of what that entails, even for one placement, for one student, for six weeks.

Another teacher also emphasised the extent of the school's responsibilities:

> Many of the universities now, even if you take on a PGCE or a BEd, it is the school that has to monitor and mentor them, and it is only when their final assessments come or if you have a major problem that you get in touch with the college or the tutor.

Certainly, student teachers on HE programmes were strongly appreciative of the role of schools. As one commented:

> Within the school setting you gain great confidence and good behavioural management techniques, which you just can't get through lectures or reading books. You need hands on experience to be able to develop your own strategies.

Heads of Faculties/Schools of Education can also be found highlighting the theme of the top-down influence of the TTA and OfSTED, although recent improvements in this respect are also noted by them (of course, highlighting that improvements are only recent serves to imply how much concern there was about previous practices).

One Head of Faculty/School was unambiguous in condemning the previous OfSTED inspection system: 'I mean the inspection regime was appalling, it was a complete waste of money it wasn't necessary and it wasn't inspection it was policing.'

However, other comments highlight recent improvements with OfSTED:

> I think they are consulting quite a lot, the whole inspection regime has shifted quite considerably. We have just been through secondary inspections where I was discussing with them areas of weakness that I needed to strengthen, and it was difficult to strengthen because of X, Y and Z. In the past regime, you couldn't say that you had to hide everything.

We also found similar, more positive comments about recent developments with the TTA:

> In terms of the agency and how it manages its constituency and relates to them it is infinitely better, and the people down there at the agency are more human and will listen and you can talk to them. Always in the past everything was punitive and a penalty not a reward. It is not quite like that now. It is about serving the sector.

A final comment was positive about recent developments both with OfSTED and the TTA:

> I would say that I think the OfSTED methodology is lighter than it was, and I think the responsiveness of the TTA to looking at quality issues is now more open than it was, and less punitive than it was, whilst still asking providers, and expecting providers to strive for quality. But I think there is a move to be more supportive and potentially to involve teacher educators more in looking at innovative solutions to the training of teachers. So I think there has been a shift. How much that shift is rhetoric at the moment, and how much there really is scope for grass roots innovation within a framework that is still strongly bounded, I am not sure I can say.

As might be expected, HE tutors more generally were perhaps taking longer to recognise any changes in OfSTED and TTA approaches. The following comment is perhaps typical of HE tutors' continuing sense of grievance with OfSTED approaches:

> I'm quite surprised that we haven't lost more teacher training providers than we already have, because of the way the system has been made to work. The inspection regime is actually quite oppressive, and it does make provision very precarious. We've lost a couple of providers because of inspection reports, and I'm not convinced they were bad providers, they were just unlucky.

School staff involved in partnership with universities also seem aware of the 'top-down' pressures from central government and agencies on the work of HE partnerships. As one commented,

> I don't think the partnership at the moment is founded in a real dialogue between higher education and the schools themselves. I think it is still too susceptible to be driven by the demands of the DfES or the TTA.

On the theme of HE concern over the potential for partnership developments to destabilise and marginalise the role of higher education within partnership, we find HE tutors, perhaps unsurprisingly, particularly critical of school-based approaches which may exclude HE completely. For example, one commented:

> It concerns me that, sometimes in the rhetoric, our relationship gets lost when we are talking about the school-based courses and there are some models that are starting to develop that do not involve HEIs at all and I think that is quite worrying.

We also find HE staff, especially Heads of Faculties/Schools, emphasising the theme of the marginalisation of LEAs within partnership. When asked to identify what the role of LEAs in partnership is, one Head of Faculty/School answered 'nothing, not in initial teacher training, I mean they [LEAs] have induction arrangements following the post qualification year but nothing for initial teacher training'.

More generally, we also found clear evidence of Heads of Faculties/Schools emphasising the need for universities to exercise the strong role in preserving appropriate approaches, as associated with the HEI-led model of partnership. As one commented, 'If you look at the data in reports on partnership, the actual extent of transfer of functions is much more limited, that's why we talk about HEI-led partnerships dominating.'

We also found evidence of HE tutors recognising that HEI-led approaches remained necessary to give overall intellectual leadership to partnerships, given that schoolteachers within partnerships have not really been able to develop that role. As one HE tutor explained:

> It's difficult to find school mentors of a research type really, they can go to courses, come to courses, we can put on INSETS for them, they come to our training days and we talk through an issue, but that can only remain in a sense a superficial level or an interest-provoking level and direction-pointing level, the teacher as researcher is few and far between, it's very difficult for them.

As another summarised, HE staff provided the overall leadership within partnership:

We provide a focus for the partners in the partnership, we are as it were the nucleus of the partnership for them to come together and share ideas as well, but we're doing the research and we are duty bound to keep up with these things and then share, so in that respect we're cascading and filtering, we're gathering ideas, we're pointing in directions, saying, 'Have you seen this? Have you seen that?'

Finally, Heads of Faculties/Schools of Education are very critical of the weaknesses within the 'non-partnership' approaches of GTPs/SCITTs. As one emphasised:

School-centred schemes, or indeed the GTP, I think it has been proven that those initiatives haven't actually delivered the substantial quality which the rest of the sector has managed to do.

This criticism of GTPs/SCITTs can also be found from HE tutors more generally. One tutor described the additionality in the PGCE experience of one of their own PGCE students who had transferred from a GTP:

If she had stayed in the school on the GTP she wouldn't have had anything like the input that she gained from being on a PGCE course, the tutors were giving her huge amounts of information and pedagogical understanding that simply wouldn't have been available to her on a GTP route.

Another HE tutor stressed the greater breadth of development experienced by student teachers on HE routes, as compared to school-based routes:

If a person were to actually graduate and then move from the school in which they have been trained, often what they only have is, this is the way in which we do it in our school and it is that rigorous intellectual challenge they miss. One would hope on a degree course that you had engaged in that wider reading. The other thing that they miss is the collaboration with their peers.

By implication, the limitations of the GTP/SCITT approaches are also evident in the views of students undertaking these routes. One student highlighted the frustrations of not being able to engage in broader discussion within his school-based route:

Because it took me a long time to understand what are these different things that we have got to do, and obviously the teachers are busy, they are very busy people, I used to say to them, can you sit down with me but it was the time factor to sit down with me and go through things and I thought, well if this was done on a wider scale at university where we could all break this down, it would be much more practical.

Even school staff who are themselves involved in GTPs/SCITTs highlighted the limitations which they encountered within these school-based routes. One teacher responsible for ITT within their school simply stressed the lack of time resource: 'I have got 30 people going through some form of training programme, it kills me. Because it takes a huge amount of time.' Another such ITT coordinator was self-critical of the ability of their school-based programme to provide staff with the knowledge and skills to support student teachers fully and appropriately:

> Do the mentors know enough themselves? Sometimes, does the person who has taught for twenty years, are they up to date? You know, you cut yourself off, you've taught for twenty years doing the same thing, you just stick to what you know, and do they know enough about the English national curriculum to pass it on to their trainee?

A deputy head teacher conceded that there were issues of consistency across school-based programmes, when staff often concentrated exclusively on the training approaches with their own school: 'I wouldn't know about another GT from another school, what their files look like.'

Finally, a school-based ITT coordinator involved in EBRs emphasised that an HE input was still required in ITT: 'I need the university sometimes to step in for the subject. Subject knowledge certainly needs to be put in, the subject knowledge in the school is not enough sometimes.'

Partnership developments in Scotland and England – discussion and conclusions

Our review of partnership developments in Scotland and England tends to suggest a clear case for divergence rather than convergence between the two systems. Of course, both Scotland and England's ITET systems have recognised the importance of addressing the international theme of strengthening partnership between HEIs and schools. However, while there have been clear moves forward in England from the HEI-based/integration/duplication model of partnership, this has not been the case in Scotland. Even if the English system has not achieved widely the fully collaborative model of ITET partnership, it has moved significantly beyond duplication into complementary/separatist models. Certainly one response from English HEIs may have been to then reassert HEI-led models as a way of preventing complete separation between the HEI and other stakeholders, and as a way of preserving a contribution for HEIs within partnership.

On the other hand, the situation in Scotland is clearly that attempts to move forward significantly beyond duplication of roles and responsibilities between HEIs and school partners have not yet succeeded. The Scottish schoolteaching profession resisted the attempt in the Mentor Teacher

Initiative of the early/mid 1990s to increase school staff responsibilities within ITE. Little specific evidence has emerged subsequently of significantly increased willingness of the Scottish schoolteaching profession to move forward on this issue. In comparison to England, this could be interpreted as a less mature acceptance by the Scottish schoolteaching profession of its role within ITET partnership. In turn, this could be linked to the perceived conservatism of the teaching profession within the GTCS, suggesting that Scottish school teachers are simply resisting the assumption of greater responsibility for ITE. On the other hand, it may be possible to argue that the Scottish schoolteaching profession's reluctance to assume a greater role within ITE reflects its respect for the role of higher education with ITE. These issues in Scotland may suggest significant divergence with England.

Of course, another explanation for divergence between Scotland and England could be linked to government policy. Particularly since the Labour election victory of 1997, and Scottish devolution since 1999, it can be argued that Scottish Ministers and administrations have not been motivated to place sufficiently high on the political agenda the actions needed to address underlying issues of partnership. This appears to contrast to the much more determined intrusions of Westminster administrations into English ITT. Such contrasts may be linked to Scottish Ministers having stronger 'pro-teacher' tendencies, and less willingness to force change upon Scottish teachers' conservative attitudes to ITE. The contrasts with England may even be linked to a stronger respect from Scottish Ministers and their advisers for the higher education role within Scottish ITE, and less desire of these Ministers and their advisers to pursue the 'technical rationalist' agenda for ITET espoused by their English counterparts. We will return to these themes in Chapter 5.

Notes

1. See Brisard et al. (2005) where partnership developments in Scotland and England are placed in the wider context of the rest of the UK, and in international contexts, including detailed consideration of examples from Australia, New Zealand and the United States.
2. As indicated in Chapter 2, the Induction scheme replaced the previous two-year minimum probationary period with a guaranteed one-year induction post for all qualified Scottish ITE graduates. The Scottish Executive contributes to local authority funding to provide enhanced staffing in schools so that induction teachers do not teach a full teaching week, and are secured support time from other experienced staff.
3. Interviews with Scottish school staff claimed three specific additional pressures: conflicting pressures on schools to raise pupils' attainment; aspects of promoted post restructuring, especially some reduction in subject-specific head of department posts in secondary schools; the competing demands of providing increased support to induction probationers (see Brisard et al., 2006, pp. 61–2).

Chapter 5

CONVERGENCE OR DIVERGENCE?

Introduction

In this chapter we review the main insights gained from the 'Convergence or Divergence' study and then reflect on the key themes in ITET that have emerged (and which were outlined in Chapter 1). We then examine what we have learned about the current state of the relationship between education policy making in Scotland and England. In conclusion we offer some broader reflections concerning teacher education and society more generally.

Scotland and England present two different policy contexts for education and teaching. As we indicated earlier, the education systems of the two countries have been distinctive throughout history, including the period from 1707 to 1999, during which the Act of Union set out the fundamental settlement of governance within Britain. Since the creation of the Scottish Parliament and the Scottish Executive, there has been even greater formal devolution of policy in many (though not all) areas of domestic policy, including education. The whole of the UK is of course part of the European Community where, since the 1990s, a European dimension in education and some degree of commonality in teacher education were being sought as part of the community's wider political and social strategy (Galton and Moon, 1994, p. 12). The EU's involvement in education and commitment to promoting quality of provision across member states was officially legitimated by Article 126 of the Maastricht Treaty, and together with the OECD, it is seen as playing an influential role in disseminating supranational policy discourses and standardised policy solutions to individual member states (Nóvoa, 2000; Henry *et al.*, 2001; Lawn and Lingard, 2002). Thus we might expect to see some degree of 'harmonisation' at the level of UK countries' policy rhetoric, general objectives and modes of governance of education and teacher education.

Unity and diversity?

This study has revealed a number of apparent similarities as well as a number of differences in the approaches taken to preparing people to teach in each country. We see a pattern therefore where there is evidence of both convergence and divergence. In the small-scale system that is Scottish ITE, it is perhaps not surprising that we have seen much greater homogeneity than is the case in the much larger English system. In Chapter 1 we foreshadowed the possibility that there is greater diversity within England than there is between Scotland and England. However, even to the extent that the diversity of approaches in England represents a divergence from the similarity of provision in Scotland, we can also see convergence in the adoption of and adherence to a set of explicit standards or competences in both countries. The question of convergence or divergence therefore is a complex one and in order to move towards some conclusions from this study, we should perhaps review what we have found by using our insights to question some of the common assumptions that are made about the comparison between the two countries, along the following lines:

Is the Scottish system actually as homogeneous as it appears?

What are the effects of the heterogeneity of the English system?

What is the significance of the apparent (relative) unity and diversity of the two systems?

Is the Scottish system as homogeneous as it appears?

In Scotland all ITE provision is offered by just seven institutions, all of them universities. There are just two major routes of entry, the four-year BEd (mainly primary) and the one-year PGDE (both primary and secondary). But there is also the concurrent degree for secondary (BA/BSc with TQ), mainly provided by just one university (Stirling). It is certainly true that the seven providers have some differences in their institutional histories and that at each institution the courses do have different structures and we now see part-time provision and distance learning being encouraged.[1]

However, in spite of these differences, all provision is accredited by the GTCS on behalf of the Scottish Executive against a common set of criteria and inspection by HMIE is carried out through 'aspect review' covering all seven providers, without published reports on individual providers or courses.

Against this backdrop, we may summarise those factors that do indicate homogeneity and contrast them with those that indicate heterogeneity. Among the factors that support the homogeneity proposition is the evidence that all providers are working hard to address the post-merger university imperatives on staff deployment which include meeting research output requirements, reviewing 'school visit' tutor roles etc. (or, in the case of Stirling, further developing these university approaches). Stakeholders generally have also identified issues concerning the conservatism of the GTCS. Finally, we have observed the very limited role of local authorities to date.

On the other hand, it is the case that in the past the Scottish Office Education Department had more formal (uniform) control over the old colleges of education than SEED exerts over the university departments, where provision is now managed by seven freestanding institutions (although the 'controlled subject' status of initial teacher education is still a 'homogenising' constraint). Indeed there are some variations in what the various universities have added to the 'merging college of education' in terms of existing education departments/units etc. There are also variations in the details of approaches to staffing across different Faculties/Schools and there are varying levels of collaboration with other disciplines (e.g. the Institute of Education at Stirling is part of a larger Faculty and at Dundee there is the Faculty of Education and Social Work). A number of claims are made for distinctive aspects of programmes, including new flexible ITE programmes. Some university providers are more critical of benchmarks/competences than others.

When we balance these two lists against each other and also reflect on what we have learned about professional knowledge (Chapter 3) and partnership (Chapter 4), what can we conclude about ITE in Scotland at the beginning of the twenty-first century? Scotland provides a model of initial teacher education that is based on a set of widely shared commitments to teaching as a respected profession, preparation for which must be firmly based within the academy. The governance and management of provision is the responsibility of a nexus of stakeholders, each of which appears to respect the role of the others. While there are tensions in the system from time to time, it is more usually characterised by consensus and mutual respect. This may be the key reason that ITE in Scotland has not been through the series of upheavals experienced in England, where central government has not been reticent in instigating change and has imposed an increasingly centralist mode of management. So, while the model of the teacher in Scotland is someone who is well educated in the theory and practice of education and someone who has been prepared to adopt inclusive and anti-discriminatory approaches (as demonstrated through Chapter 3), s/he has not necessarily been prepared through an integrated model of partnership between schools and the academy (as indicated in Chapter 4). The structural shifts which would be required to ensure the adoption by teachers and schools of a more formal responsibility for the preparation of new teachers have not been achieved in Scotland. There may be many examples of 'good practice' across Scotland, involving serving teachers in a number of different ways in ITE provision, but the systematic approach that would ensure all beginning teachers work with serving teachers who are trained to support them in an effective way has not yet been achieved.

What are the effects of the heterogeneity of the English system?

Before answering this question, we should remind ourselves of what we mean by the heterogeneity of the English system. There are three key aspects to diversity:

- different routes – from employment-based routes that lead only to QTS to academic awards such as PGCE and BEd that incorporate QTS;
- different providers – individual schools and school consortia, independent providers, universities, colleges;
- different partnerships – approaches involving little or no HEI involvement through to partnerships led and managed by HEIs.

However, all of the provision is working to the same set of standards and all of it is managed and controlled by the TTA (now TDA) and by OfSTED.

Here again, in answer to the main question it may be helpful to contrast a number of problems – real or potential – with heterogeneity, with a number of benefits. Our study has suggested that problems with heterogeneity include the fact that experiences in training may vary considerably, that varied conceptions of teaching may underlie the different approaches, that it may be very difficult to ensure consistency in the school-based experience of new teachers as well as in their assessment against the standards, and there may be a lack of uniform quality assurance procedures because of the very different contexts in which training takes place. All of these factors may lead to variable quality of NQTs and each different approach may imply or require different subsequent approaches to career development. Overall, there is a reduced involvement of HE (this is not seen as a problem by all) which may in turn lead to a reduction of perceived status of teachers. Finally, there is a lack of professional control by a teachers' body and it seems that teacher professionalism in England can now legitimately range from 'restricted' to 'extended' forms (as discussed at the end of Chapter 3).

By contrast, what may be the benefits of such heterogeneity? The greater diversity of entry routes may mean that a broader range of entrants may be attracted into teaching – including some who would not otherwise join teaching. Through creating targeted incentivisation it may be possible to create a more responsive supply of teachers (to where they are needed and for what subjects or age phases). It has also been frequently asserted by politicians that a more competitive environment will lead to improvements in quality (although it is difficult to substantiate this linkage). There may be greater relevance in the training experience for many trainees and a range of modes of involvement in initial training may be provided for schools and teachers, thus creating additional professional development opportunities. It does appear that greater diversity creates the potential for involvement by a greater range of stakeholders.

However, a significant overarching question arises in England: is there an emergent fracturing of a common collective professional identity of teachers in England? The government has always claimed that through assessing all beginning teachers against the same set of standards, whatever their route of entry, it may be ensured that a common level of competence has been reached. Such an argument is based on the beliefs that:

- all that is important about being a teacher can be defined through a set of standards;
- different approaches towards the achievement of the standards are not only not significant at the point of assessment but will not become significant thereafter (OfSTED reports question the basis for future development provided by EBRs);
- so long as the 'baseline' of the standards is achieved it does not matter whether it is exceeded or not.

All of these points are facilitated by the fact that the teaching profession is increasingly 'managed', with decisions about what to teach, how to teach it and how to assess children being made at school and national level rather than by individual teachers themselves (Furlong, 2005, p. 120).

The nature of the identity of the teaching profession is becoming even more complex and potentially even more differentiated as the roles – and numbers – of support staff are significantly increasing. In England the Higher Level Teaching Assistants (HLTAs) are now legally entitled to supervise classes when teachers are not available. Our study leads us to conclude that there may be a serious division between those entering the workforce according to their route of entry. We cannot avoid invoking the possibility that we see a return in England to the legitimation of a restricted form of professionalism for some teachers, which runs strongly counter to the consistent move towards extended professionalism that occurred during the last three decades of the twentieth century.

What is the significance of the apparent (relative) unity and diversity of the two systems?

On the basis of these comparisons, although neither the homogeneity of Scotland nor the heterogeneity of England are as consistent as might at first appear, we may nevertheless conclude that there are significant differences between ITET in Scotland and England. What is more difficult to assess is whether these differences are increasing (diverging) or decreasing (converging) at present. There is certainly historical evidence that many of the differences go back well before the recent devolution settlement. Whether, as Scottish devolution 'matures', we will see greater divergence and distinctiveness is far from clear at present. The forces towards convergence appear to be growing in strength as supra-national influences grow.

Key themes in initial teacher education and training

The five themes that we introduced in Chapter 1 may serve as a basis on which to reflect on these trends, both to summarise the present situation and as 'indicators' that may help to assess the patterns of change into the future.

Accountability

The drive for accountability has been a strong feature across public services in the Western world for many years. Its manifestation in Scottish ITE has been more collegial and supportive than the equivalent developments in English ITT which have at times been punitive and in all cases have sought to link 'performance', as judged by quality, to rewards. The latter has led to some destabilisation in the English system, which has created difficulty, especially at times of teacher shortage.

Universitisation

While the process of moving all Scottish ITE into universities was completed within the last few years, the English system has seen a significant diminution in the role of HEIs. In most other parts of Europe and the wider world, the Scottish trend appears to be more prevalent than the English. It is generally the case that in most countries there has been a commitment to improve the academic standing of teacher education.

Rationalisation

In this age of public accountability there is an underlying drive for economic efficiency. In neither country has this necessarily been very visible in the recent past, although the universitisation drive in Scotland may be partly seen in this way. It has for example led to a significant reduction in the number of HEIs that the Scottish Funding Council has had to support. In England, on the other hand, the diversification agenda is difficult to connect with rationalisation. Some of the new routes, including the employment-based routes, may help to address immediate staff shortages in some schools, but they are not necessarily more cost effective, since they involve the payment of a salary from the outset, as well as supporting a training programme.

Professionalisation

The 'modernisation' of the teaching profession is a widespread, indeed global, phenomenon (Smyth *et al.*, 2000). Nevertheless, as we have seen in this study it can take very different forms. Although ITET is rarely at the forefront of programmes of reform in the teaching profession (and this feature is something of a curiosity), wider reforms do eventually have an impact. The process of review and reform of teachers and teaching in Scotland, instigated through the establishment of the McCrone Committee, did lead to the two stage review of ITE that we have described and this is having some influence on the development of ITE in Scotland. However, it has not led to the kind of 'fracturing' of the teaching profession that has been under way in England and may be traced back to a range of initiatives including the 1998 Green Paper (DfEE, 1998b) and the more recent policy of 'workload agreement' and 'workforce reform'.[2] These policies, together with the diverse range of entry routes into the profession, do reveal a very different take on teacher professionalism to that which is prevalent in Scotland.

Partnership

The theme of partnership is one that is especially telling in analysing teacher education systems. Indeed we have devoted a whole chapter to this theme because it is so significant. We have dwelt on the Scottish efforts to bring about a greater engagement of schools and teachers in the process of ITE and found some explanations for the ways in which these efforts have faltered. This does stand in stark contrast with the English scenario where a very dogmatic and determined Conservative government pushed through major reforms in the early 1990s that do appear to have had a number of significant effects that are now (more or less) embedded in the system. While there are still a number of concerns about these developments (for example, HEIs still tend to complain that they have all the responsibility for securing effective partnerships and that they cannot easily influence 'quality' in schools), it is the case that most entrants into teaching in England now experience contact with trained classroom-based staff. On the other hand, when they *only* have contact with such staff and have none with HEI staff during their training, as can happen on employment-based routes, it is actually inappropriate to refer to this as a partnership, at least in the conventional sense.

These five themes do help us to characterise some of the respective distinctiveness of the two ITET systems. We would suggest that researchers with an interest in these matters should keep these themes under review during the years ahead. This is likely to give valuable insights into the respective policy trajectories in each country (or indeed in any other country).

Policy development in Scotland and England

In Chapter 1 we suggested that this study might offer some opportunity to assess the relative significance of three contrasting ways of understanding the relationship between Scottish and English policy making in education. These three models might be described as respectively 'parallel development', 'UK specific' and 'globalisation'. The parallel development thesis is one that emerges most strongly when a longer historical view is taken and one can trace similarities in policy that perhaps do not emerge at exactly the same time, but nevertheless emerge within the same wider time frame. Although the development of standards and benchmark/competences use different language, they do have something of the same underlying rationale and might be seen as an example of this kind of parallel development (but see also below).

The UK specific model relates to the tensions and power struggles that have existed between the constituent parts of the UK. It is perhaps not surprising that, in the wake of the 1999 devolution programme across the UK, this does appear to have been of greater significance during the recent past and in particular, during the time that data was being gathered for this study. In Scotland this had perhaps best been demonstrated by a reluctance to use English developments as a model. While prior to devolution there was often a concerted effort to resist the encroachment into Scotland of what were perceived as extreme policies being enacted in England, since devolution there has been a greater determination to be self-sufficient and/or to look to other parts of Europe and elsewhere, as appropriate, for influence or inspiration. We have seen the way in which such a perspective may have influenced the partnership agenda in a confusing manner.

On the other hand, in Scotland's desire to maintain the good reputation of its education system as a whole and of the high quality of its teachers, as widely perceived, the forces of 'globalisation' are having some impact. The development of standards or competences/benchmarks can be seen as a result of global forces, at least as much as of parallel developments in the two countries. The awareness of increased global competitiveness in education systems has been heightened by the international comparisons of pupil achievement such as PISA, TIMSS and PIRLS.[3] Part of the English response to these pressures has been to place great emphasis on literacy and numeracy teaching and training and this has now become a central part of ITT. In Scotland, there has not been the equivalent imposition of particular pedagogies, and reform both in the school curriculum and in ITE has tended to be more gradual and holistic. However, if political concerns about levels of attainment grow in Scotland as they did in England, this is likely to have a not insignificant impact on ITE, at least in due course.

So, to summarise our review of these three approaches, we can see all three models of policy development continuing to have some relevance, although the UK specific and the global dimensions appear to have assumed greater significance in recent times. If we now step back from the immediate questions of similarities and differences that we have identified between the two systems, are there some more general insights that may be drawn from our comparison? We would make three interconnecting points that may be summarised as below.

First, we refer to teacher education and the cultural significance of education. A close connection has emerged in this study between the conditions under which teachers join the profession and the wider standing of education within society. Although teachers in Scotland have felt under attack in recent times, they have experienced nothing like the vitriol of the 'discourse of derision' encountered by teachers in England (Ball, 1990). Where teachers and teaching are more highly respected, their preparation is more carefully defended and defined.

Second, in relation to teacher identity and national identity, we would suggest that the identity of teachers in a small nation, that has experienced centuries of, if not subjugation, then at least domination, by a larger more powerful neighbour, tends to be more closely and explicitly aligned to a national view of the purposes of education. Following increased independence post-devolution, it seems that teachers' professional identity may be even more closely connected with the development of the wider national identity, as a small European nation, seeking to 'punch above its weight' (as epitomised by the Scottish Executive slogan 'Building the best small country in the world').

Third, on the question of differing definitions of teacher professionalism, it appears that a society that values teachers more highly may be less likely to proceed down a technical-rationalist model of teacher professionalism that seeks to reduce teaching performance to a list of observable skills. The managerialism of Third Way New Labour politics apparently has far less purchase in a culture and society which still seems to attribute certain values to scholarship and learning – indeed where the idea of the democratic intellect may continue to be a meaningful cultural concept (Davie, 1961).[4]

Conclusion

The 'discourse of derision' (Ball, 1990) which was so directly targeted at teachers in England in the 1980s was not a feature of Scottish popular discourse over the same period. Many aspects of Conservative education policy during the Thatcher and Major years were strongly resisted in Scotland (see Paterson, 2003). Key features of the Scottish policy community are very different from England – including the long history of a General Teaching Council, the strong domination of one teachers' union, the absence of a TTA or an OfSTED, which can explain how Scotland was able to moderate or mediate prior to devolution some of the changes proposed south of the border.

We have also drawn attention to the underlying economic features – especially the issues of supply and demand – that are often the invisible driver of policy and practice in ITET. Such factors will continue to interact with the nature of provision in both countries and may give rise to instability in either system. The rapid expansion of teacher supply in Scotland resulting more from policy initiatives than from demography has not been without challenges for providers, albeit in a very different way from the difficulties faced in England as numbers have fluctuated and new routes have come on stream.

Indeed, the situation regarding initial teacher education/training is not static in either country. In England current developments include the emergence of a much less caustic relationship between the university providers and the TTA, continuing attention to developments in partnership, including partnership promotion and the identification of 'training schools' (Brisard *et al.*, 2005). In Scotland, the report of the Executive's 'second stage review' of ITE was eventually published in June 2005. It indeed commented on the roles of and relationships of stakeholders in the initial education of teachers, although the word partnership was avoided. Given the almost fierce commitment to an all graduate profession, it was no surprise that the university-led element of the process was not significantly undermined.

Olssen *et al.* (2004) associate managerialism in professional work with the neoliberalism which has permeated public policy in the 'advanced' societies. Teachers – both in schools and in post-compulsory education – have become part of a high accountability, low trust society. They suggest that:

> The specification of objectives, performance reviews and other management techniques may encourage teachers to behave in ways that are antithetical to certain fundamental educational values such as altruism, intellectual independence and imagination. Moreover, we argue that the restoration of a culture of trust and professional accountability within all educational institutions is a necessary prerequisite for the maintenance of a robust and prosperous democratic society. (p. 197)

Certainly, by this reading, the differences which we have detected above between English and Scottish policies would seem only to reflect different gradations of neoliberal influence, as indicated by the common commitment to a standards-based approach to the technical skills of teaching. Certainly such 'fundamental educational values' are more prevalent in Scottish policy discourse and one can at least identify them there, whereas in the English discourse surrounding initial teacher training at least, they have been obscured, if not obliterated. However, even in Scotland, such values would appear to be under some pressure, particularly the values of independence and imagination. What Ball (2005) has called 'authentic teacher professionalism' is at least threatened in Scotland (see Ozga, 2004), but not (yet) extinguished, as he suggests it is in England and elsewhere.

This has been an ambitious study, during which a large amount of data has been gathered that we hope to continue mining for further insights. What we have attempted to do in this book is to offer a context-sensitive analysis of policy and practice in ITET in the two countries. The underlying purpose has been to examine broad questions about the nature of contemporary education policy making within the UK, to explore some of the linkages between ITET systems and the nation and to explore all of these matters in the specific context of a period of significant political change. Indeed, at this stage in our studies, it would seem that there is a danger of attaching too much significance to the specific Scottish devolution settlement. What we have found is considerable continuity, especially in Scotland, influenced by cultural and institutional histories. While there are significant divergences between the two countries, there are also simultaneously significant convergences, several of which are responses to wider influences.

Not only is there scope for further work on our own data, there is a great need for further work of this kind that can draw in the other parts of the UK (Wales and Northern Ireland), as well as perhaps extending into other domains, both English speaking (such as the Republic of Ireland) and non-English speaking (other European countries).

Notes

1. Again, at one university (Aberdeen), we have recently seen a fresh initiative in the form of 'Scottish Teachers for a New Era', funded by the Scottish Executive and a private philanthropic fund, the Hunter Foundation. There are some aspirations that this development will influence provision across the whole of Scotland, but at present it is not entirely clear how this might happen.
2. A national agreement on *Raising standards and tackling workload* was signed by government, employers and school workforce unions on 15 January 2003, for details see: www.remodelling.org/Home/remodelling/nationalagreement/introduction. aspx (accessed 28 April 2006).
3. PIRLS: The Progress in International Reading Literacy Study; PISA: The Programme for International Student Assessment; TIMMS: the Trends in International Mathematics and Science Study.
4. The idea of the democratic intellect is a concept not without its critics, including those who say it is another element of the romanticised mythology of Scottish superiority (see Humes and Bryce, 2003).

BIBLIOGRAPHY

Adick, C. (2002) 'Demanded and feared: transnational convergencies in national educational systems and their (expectable) effects', *European Educational Research Journal*, Vol. 1, No. 2, pp. 214–33

Alexander, R. (2000) *Culture and Pedagogy*, Oxford: Blackwell

Alexander, R. (2001) 'Border crossing, towards a comparative pedagogy', *Comparative Education*, Vol. 37, No. 4, pp. 507–23

Anderson, R. (1995) *Education and the Scottish People, 1750–1918*, Oxford: Clarendon

Arnove, R. (1999) 'Reframing comparative education, the dialectic of the global and the local', in Arnove, R. F. and Torres, C. A. (1999) *Comparative Education, the Dialectic of the Global and the Local*, New York and Oxford: Rowman & Littlefield, pp. 1–23

Arnove, R. F. and Torres, C. A. (1999) *Comparative Education, the Dialectic of the Global and the Local*, New York and Oxford: Rowman & Littlefield

Baistow, K. (2000) 'Cross-national research: what can we learn from inter-country comparisons?' *Social Work in Europe*, Vol. 7, No. 3, pp. 8–13

Ball, S. (1990) *Politics and Policy Making in Education*, London: Routledge

Ball, S. (1994) *Education Reform: A Critical and Post-Structural Approach*, Buckingham: Open University Press

Ball, S. (1998) 'Big policies/small world: an introduction to international perspectives in educational policy', *Comparative Education*, Vol. 34, No. 2, pp. 117–31

Ball, S. (2005) 'Education reform as social barberism: economism and the end of authenticity', SERA Lecture 2005, *Scottish Educational Review*, Vol. 37, No. 1, pp. 4–16

Beck, U. (2000) *What is Globalization?*, Cambridge: Polity Press

Bertaux, D. (ed.) (1981) *Biography and Society*, Beverly Hills: Sage

Blumer, H. (1971) 'Sociological implications of the thought of George Herbert Mead', in Cosin, B. R., Dale, I. R., Esland, G. M., MacKinnon, D. and Swift, D. F. (eds) (1971) *School and Society*, London: Routledge & Kegan Paul /Open University Press, pp. 11–17

Bowe, R. and Ball, S. with Gold, A. (1992) *Reforming Education and Changing Schools: Case Studies in Policy Sociology*, London: Routledge

Brisard, E. and Menter, I. (2004) 'Compulsory education in the United Kingdom', in Matheson, D. (ed.) (2004) *An Introduction to the Study of Education*, 2nd edn, London: David Fulton Publishers, pp. 184–212

Brisard, E., Menter, I. and Smith, I. (2005) *Models of Partnership in Initial Teacher Education*. Full Report of a Systematic Literature Review Commissioned by the General Teaching Council for Scotland. GTCS Research, Research Publication No. 2, Edinburgh: GTCS

Brisard, E., Menter, I. and Smith, I. (2006) 'Discourses of partnership in initial teacher education in Scotland : current configurations and tensions', *European Journal of Teacher Education*, Vol. 29, No. 1, pp. 49–66

Bryce, T. and Humes, W. (eds) (2003) *Scottish Education, second edn post-devolution*, Edinburgh: Edinburgh University Press

Buchberger, F., Campos, B. P., Kallos, D. and Stephenson, J. (eds) (2000) *Green Paper on Teacher Education in Europe: High Quality Teacher Education for High Quality Education and Training*, Thematic Network for Teacher Education in Europe (TNTEE): Umea University, Sweden

Cameron-Jones, M. (1995) 'Permanence, policy and partnership in teacher education', in Kirk, G. (ed.) (1995) *Moray House and Change in Higher Education*, Edinburgh: Scottish Academic Press, pp. 21–35

Cameron-Jones, M. and O'Hara, P. (1993) *The Scottish Pilot PGCE (Secondary) course 1992–1993*, Edinburgh: Moray House Institute / Heriot-Watt University

Cameron-Jones, M. and O'Hara, P. (1994a) 'Pressures on the curriculum of teacher education', *Scottish Educational Review*, Vol. 26, No. 2, pp. 134–42

Cameron-Jones, M. and O'Hara, P. (1994b) *The Second Year (1993–1994) of the Scottish Pilot PGCE (Secondary) Course*, Edinburgh: Moray House Institute / Heriot-Watt University

Campbell, A., McNamara, O. and Gilroy, P. (2004) *Practitioner Research and Professional Development in Education*, London: Paul Chapman

Carr, D. (1993) 'Questions of competence', *British Journal of Educational Studies*, Vol. 41 No. 3, pp. 253–71

Crossley, M. and Watson, K. (2003) *Comparative and International Research in Education – Globalisation, Context and Difference*, London: RoutledgeFalmer

Crouch, C. (2003) *Commercialisation or Citizenship: Education Policy and the Future of Public Services*, London: Fabian Society

Dale, R. (1989) *The State and Education Policy*, Milton Keynes: Open University Press

Dale, R. (1999) 'Specifying globalisation effects on national policy: a focus on the mechanisms', *Journal of Educational Policy*, Vol. 14, No. 1, pp. 1–17

Davie, G. (1961) *The Democratic Intellect*, Edinburgh: Edinburgh University Press

Day, C. (1999) *Developing Teachers – the Challenges of Lifelong Learning*, London: RoutledgeFalmer

Deloitte & Touche (1999) *Costs of Partnership in Initial Teacher Education*, final report to the Scottish Office Education and Industry Department and the General Teaching Council, Edinburgh: Deloitte Touche

Deloitte & Touche (2001) *Report of the 'First Stage' Review of Initial Teacher Education, Independent review of the Scottish ITE sector*, Edinburgh: Scottish Executive

Dent, H.C. (1977) *The Training of Teachers in England and Wales: 1800–1975*, London: Hodder & Stoughton

Department for Education (DfE) (1992) *Initial Teacher Training (secondary phase)* (Circular 9/92), London: DfE

Department for Education (DfE) (1993) *The Initial Training of Primary School Teachers: New Criteria for Courses* (Circular 14/93), London: DfE

Department for Education and Employment (DfEE) (1997) *Teaching: High Status, High Standards* (Circular 10/97), London: DfEE

Department for Education and Employment (DfEE) (1998a) *Requirements for Courses of Initial Teacher Training* (Circular 4/98), London: DfEE

Department for Education and Employment (DfEE) (1998b) *Teachers: Meeting the Challenge of Change*, Green Paper, London: DfEE

Department for Education and Employment (DfEE) and Qualifications and Curriculum Authority (QCA) (1999) *The National Curriculum: Handbook for Primary Teachers in England and Handbook for Secondary Teachers in England*, London: DfEE and QCA

Department of Education and Science (DES) (1984) *Initial Teacher Training: Approval of Courses* (Circular 3/84), London: DES

Department of Education and Science (DES) (1989) *Initial Teacher Training: Approval of Courses* (Circular 24/89), London: DES

Edwards, A., Gilroy, P. and Hartley, D. (2002) *Rethinking Teacher Education – Collaborative Responses to Uncertainty*, London: RoutledgeFalmer

Elder, R. and Kwiatkowski, H. (1993) *Partnership in Initial Teacher Education*, Edinburgh: Scottish Education Department/General Teaching Council Scotland Research Report

Epstein, E. (1992) 'The problematic meaning of "comparison" in comparative education', in Schriever, J. and Holmes, B. (eds) (1992) *Theories and Methods in Comparative Education*, third edn, Komparatistische Bibliothek, Franfurt am Main: Peter Lang, pp. 3–23

Eraut, M. (1994) *Developing Professional Knowledge and Competence*, London: RoutledgeFalmer
Eurydice (2002) *The Teaching Profession in Europe: Profiles, Trends and Concerns: Report I: Initial Training and Transition to Working Life. General Lower Secondary*. Key topics in Education in Europe, Vol. 3, Brussels: Eurydice European Unit
Fielding, M. (2001) (ed.) *Taking Education Really Seriously: Four Years' Hard Labour*, London: RoutledgeFalmer
Furlong, J. (1995) 'The limits of competence: a cautionary note on Circular 9/92', in Kerry, T. and Shelton Mayes, A. (eds) (1995) *Issues in Mentoring*, London: Routledge, pp. 225–31
Furlong, J. (2000) *Higher Education and the New Professionalism of Teachers*, London: Universities UK and SCOP
Furlong, J. (2005) 'New Labour and teacher education: the end of an era', *Oxford Review of Education*, Vol. 31, No. 1, pp. 119–34
Furlong, J., Barton, L., Miles, S., Whiting, C. and Whitty, G. (2000) *Teacher Education in Transition – Reforming Professionalism?*, Buckingham: Open University Press
Furlong, J., Campbell, A., Howson, J., Lewis, S. and McNamara, O. (2006) 'Partnership in English Initial Teacher Education: changing times, changing definitions, evidence from the TTA National Partnership Project', *Scottish Educational Review*, Vol. 37 (Special Edition), pp. 21–32
Galton, M. and Moon, B. (eds) (1994) *Hanbook of Teacher Training in Europe*, London: David Fulton Publishers/ Council of Europe
Gauthier, P.-L. (2002) 'La formation des enseignants dans l'Union européenne: enjeux et modèles', paper delivered at the *Conférence Internationale de l'Association Francophone d'Education Comparée*, Caen, France, 25–28 May 2002
Geertz, C. (1973) *The Interpretation of Cultures*, London: Hutchinson
General Teaching Council for Scotland (GTCS) (1997) *The Report of the Working Group on Partnership in Initial Teacher Education*, Edinburgh: GTCS
Gewirtz, S., Dickson, M. and Power, S. (2004) 'Unravelling a "spun" policy: a case study of the constructive role of "spin" in the education policy process', *Journal of Education Policy*, Vol. 19, No. 3, pp. 321–42
Gilroy, P. (1992) 'The political rape of Initial Teacher Education in England and Wales: a JET rebuttal', *Journal of Education for Teaching*, Vol. 18, No. 1, pp. 5–22
Goodson, I. (2003) *Professional Knowledge, Professional Lives*, Maidenhead: Open University Press
Grace, G. (1984) 'Urban education: policy science or critical scholarship?' in Grace, G. (ed.) (1984) *Education and the City: Theory, History and Contemporary Practice*, London: Routledge & Kegan Paul, pp. 3–59
Green, A. (1999) 'Education and globalization in Europe and East Asia: convergent and divergent trends', *Journal of Education Policy*, Vol. 14, No. 1, pp. 55–71
Green, A. (2002) *Education, Globalisation and the Role of Comparative Research*, London: Institute of Education
Henry, M., Lingard, B., Rizvi, F. and Taylor, S. (2001) *The OECD, Globalisation and Education Policy*, London: Pergamon
Hoffman, D. M. (1999) 'Culture and comparative education: towards decentering and recentering the discourse', *Comparative Education Review*, Vol. 43, No. 4, pp. 464–88
Holmes, H. (2000) *Education*, Vol. 11 of Scottish Life and Society, Edinburgh: Tuckwell Press
Humes, W. (1986) *The Leadership Class in Scottish Education*, Edinburgh: John Donald
Humes, W. (1995) 'From disciplines to competences: the changing face of professional studies in teacher education', *Education in the North*, New Series, Vol. 3, pp. 39–47
Humes, W. and Bryce, T. (2003) 'The distinctiveness of Scottish education', in Bryce, T. and Humes, W. (eds) (2003) *Scottish Education, second edn post-devolution*, Edinburgh: Edinburgh University Press, pp. 102–14

Jarvis, P. (2002) 'Globalisation, citizenship and the education of adults in contemporary european Society', *Compare*, Vol. 32, No. 1, pp. 5–19

Keating, M. (2002) 'Devolution and public policy in the United Kingdom: divergence or convergence?' in Adams, J. and Robinson, P. (eds) *Devolution in Practice: Public Policy Differences within the UK*, London: Institute for Public Policy Research, pp. 3–24

Kirk, G. (2000) *Enhancing Quality in Teacher Education*, Edinburgh: Dunedin Academic Press

Kogan, M. (1975) *Educational Policy-Making: A Study of Interest Groups and Parliament*, London: Allen and Unwin

Lave, J. and Wenger, E. (1991) *Situated Learning: Legitimate Peripheral Participation*, Cambridge: Cambridge University Press

Lawn, M. and Lingard, B. (2002) 'Constructing a European policy space in educational governance: the role of transnational policy actors', *European Educational Research Journal*, Vol. 1, No. 2, pp. 290–307

Mahony, P. and Hextall, I. (2000) 'Consultation and the management of consent: standards for Qualified Teacher status', *British Educational Research Journal*, Vol. 26, No. 3, pp. 323–42

Mahony, P. and Hextall, I. (2001) *Reconstructing Teaching*, London: RoutledgeFalmer

Mahony, P., Hextall, I. and Menter, I. (2001) 'Just Testing?: an analysis of the implementation of "skills tests" for entry into the teaching profession in England", *Journal of Education for Teaching*, Vol. 27, No. 3, pp. 221–39

Marker, W. (2000) 'Scottish teachers', in Holmes, H. (ed.) (2000) *Education*, Vol. 11 of Scottish Life and Society, Edinburgh: Tuckwell Press, pp. 273–96

McIntyre, D. (ed.) (1997) *Teacher Education Research in a New Context: The Oxford Internship Scheme*, London: Paul Chapman

McPherson, A. and Raab, C. D. (1988) *Governing Education: A Sociology of Policy Since 1945*, Edinburgh: Edinburgh University Press

Menter, I. (2002) 'Border crossing – teacher supply and retention in England and Scotland', *Scottish Educational Review*, Vol. 34, No. 1, pp. 40–50

Menter, I., Brisard, E. and Smith, I. (2006) 'Making teachers in Britain: professional knowledge for initial teacher education in England and Scotland', *Educational Philosophy and Theory*, Vol. 38, No. 3, pp. 269–86

Menter I., Mahony P. and Hextall, I. (2004) 'Ne'er the twain shall meet? The modernisation of the teaching workforce in Scotland and England', *Journal of Education Policy*, Vol. 19, No. 2, pp. 195–214

Merryfield, M. (1994) *Teacher Education in Global and International Education*, Washington, DC: American Association of Colleges for Teacher Education

Moreau, M.-P. (2005) 'Comment problématiser la comparaison internationale?', in Malet, R. and Brisard, E. (eds) (2005) *Modernisation de l'Ecole et Contextes Culturels: des polotiques aux pratiques en France et Grande-Bretagne*, Paris: L'Hartmann, pp. 91–110

Morrow, R. and Torres, C. (2000) 'The state, globalization and education policy', in Burbules, N. and Torres, C. (eds) (2000) *Globalization and Education: Critical Perspectives*, London: Routledge, pp. 27–56

Newman, J. (2001) *Modernizing Governance: New Labour, Policy and Society*, London: Sage

Nóvoa, A. (2000) 'The teaching profession in Europe: historical and sociological analysis', in Seing, E. S., Schriewer, J. and Orivel, F. (eds) (2000) *Problems and Prospects in European Education*, Westport, CN: Praeger, pp. 45–71

Office for Standards in Education (OfSTED) (2002) *The Graduate Teacher Programme. A Report from Her Majesty's Chief Inspector of Schools in England, 2000–1*, London: OfSTED

Olssen, M., Codd, J. and O'Neill, A.-M. (2004) *Education Policy – Globalization, Citizenship and Democracy*, London: Sage

Osborn, M. and McNess, E. (2005) 'Les transformations du travail des enseignants en Angleterre, en France et au Danemark', in Malet, R. and Brisard, E. (eds) (2005), *Modernisation de l'Ecole et Contextes Culturels: des polotiques aux pratiques en France et Grande-Bretagne*, Paris: L'Hartmann, pp. 173–94

Ozga, J. (2000) *Policy Research in Educational Settings*, Buckingham: Open University Press

Ozga, J. (2004) 'Modernising the education workforce: a perspective from Scotland', *Educational Review*, Vol. 57, No. 2, pp. 207–21

Ozga, J. (2005) 'L'étude comparée des politiques de l'école et du travail des enseignants', in Malet, R. and Brisard, E. (eds) (2005), *Modernisation de l'Ecole et Contextes Culturels: des polotiques aux pratiques en France et Grande-Bretagne*, Paris: L'Hartmann, pp. 31–50

Parker, S. (1997) *Reflective Teaching in the Post-Modern World*, Buckingham: Open University Press

Paterson, L. (2000) *Education and the Scottish Parliament*, Edinburgh: Dunedin Academic Press

Paterson, L. (2003) *Scottish Education in the Twentieth Century*, Edinburgh: Edinburgh University Press

Pickard, W. and Dobie, J. (2003) *The Political Context of Education after Devolution*, Edinburgh: Dunedin Academic Press

Popkewitz, T., Franklin, B. and Pereyra, M. (eds) (2001) *Cultural History and Education*, London: RoutledgeFalmer

Popkewitz, T., Lindblad, S. and Stranberg, J. (1999) *Review of Research on Education Governance and Social Integration and Exclusion*, Uppsalla Reports on Education, Vol. 36, Department of Education: Uppsalla University

Quality Assurance Agency for Higher Education (QAA) (2000) *The Standard for Initial Teacher Education in Scotland, Benchmark Information*, Gloucester: Quality Assurance Agency for Higher Education

Raffe, D. (1998) 'Does learning begin at home? The use of "home International" comparisons in UK policy-making', *Journal of Education Policy*, Vol. 13, No. 5, pp. 591–602

Raffe, D., Brannen, K., Croxford, L. and Martin, C. (1999) 'Comparing England, Scotland, Wales and Northern Ireland: the case for "home internationals" in comparative research', *Comparative Education*, Vol. 95, No. 1, pp. 9–25

Robertson, S. (2000) *A Class Act: Changing Teachers' Work, The State and Globalisation*, London: Falmer

Ryba, R. (1992) 'Common trends in teacher education in European Community countries', *Compare*, Vol. 22, No. 1, pp. 25–39

Sachs, J. (2003) *The Activist Teaching Profession*, Buckingham: Open University Press

Sander, T. (2000) 'The politics of comparing teacher education systems and teacher education policy', in Alexander, R., Osborn, M. and Phillips, D. (eds) (2000) *Learning from Comparing, Volume 2: Policy, Professionals and Development*, Oxford: Symposium Books, pp. 161–203

Schnaitmann, G. W. (ed.) (1998) *Comparison, Understanding and Teacher Education in International Perspective*, Frankfurt am Main: Peter Lang

Schon, D. (1987) *Educating the Reflective Practitioner*, San Francisco: Jossey-Bass

Scottish Executive (2000) *A Teaching Profession for the 21st Century, Volume 1 and 2: Report of the Committee of Inquiry into Professional Conditions of Service of Teachers* (The McCrone Report), Edinburgh: Scottish Executive

Scottish Executive (2001a) *A Teaching Profession for the 21st Century, Agreement reached following recommendations made in the McCrone Report* (The McCrone Settlement), Edinburgh: Scottish Executive

Scottish Executive (2001b) *First Stage Review of Initial Teacher Education: Action Plan*, Edinburgh: Scottish Executive

Scottish Executive (2003) 'Next steps in teacher training reform', press release, 22 September 2003 (online). Available from URL: www.scotland.gov.uk/News/Releases/2003/09/4153

(accessed 21 January 2005)

Scottish Executive (2005a) *Review of Initial Teacher Education Stage 2*, Edinburgh: Scottish Executive

Scottish Executive (2005b) *Review of Initial Teacher Education Stage 2, Ministerial Response*, Edinburgh: Scottish Executive

Scottish Office Education and Industry Department (SOEID) (1998) *Guidelines for Initial Teacher Education Courses in Scotland*, Edinburgh: HMSO

Smith, I., Brisard, E. and Menter, I. (2006a) 'Partnership in Initial Teacher Education in Scotland, 1990–2005: unresolved tensions', *Scottish Educational Review*, Vol. 37, Special Edition, pp. 20–31

Smith, I., Brisard, E. and Menter, I. (2006b) 'Models of partnership development in initial teacher education in the four components of the UK: recent trends and current challenges', *Journal of Education for Teaching*, Vol. 32, No. 2, pp. 147–64

Smyth, E., Gangl, M., Raffe, D., Hannan, D. F. and McCoy, S. (2001) *A Comparative Analysis of Transitions from Education to Work in Europe* (CATEWE), Final report, Dublin: Economic and Social Research Institute (ESRI)

Smyth, J., Dow, A., Hattam, R., Reid, A. and Shacklock, G. (2000) *Teachers' Work in a Globalizing Economy,* London: Falmer

Stenhouse, L. (1975) *An Introduction to Curriculum Research and Development*, London: Heinemann

Stephenson, J. (1994) 'Anatomy of a development', in Yeomans, R. and Sampson, J. (eds) (1994) *Mentorship in the Primary School*, London: Falmer, pp. 19–32

Stronach, I., Cope, P., Inglis, B. and McNally, J. (1994) 'The SOED "Competences" guidelines for initial teacher training: issues of control, performance and relevance', *Scottish Educational Review*, Vol. 26, pp. 118–33

Stronach, I., Cope, P., Inglis, B. and McNally, J. (1996) '"Competence" guidelines in Scotland for Initial Teacher Training: "supercontrol" or "superperformance"'? in Hustler, D and McIntyre, D. (eds) (1996) *Developing Competent Teachers*, London: David Fulton Publishers, pp. 72–85

Teacher Training Agency (TTA) and Department for Education and Skills (DfES) (2002) *Qualifying to Teach: Professional Standards for Qualified Teacher Status and Requirements for Initial Teacher Training* (Circular 02/02), London: TTA

Teacher Training Agency (2004a) *Qualifying to Teach: Professional Standards for Qualified Teacher Status and Requirements for Initial Teacher Training*, London: TTA

Teacher Training Agency (2004b) *Qualifying to Teach: Handbook of Guidance*, London: TTA

Thornton, K. and Munro, N. (2001) 'English "on the job" training fails Scottish test', *Times Education Supplement*, 21 December, p. 4

Tomlinson, S. (2001) *Education in a Post-Welfare Society*, Buckingham: Open University Press

Walford, G. (2001) 'Site selection within comparative case study and ethnographic research', *Compare*, Vol. 31, No. 2, pp. 151–63

Wenger, E. (2000) *Communities of Practice: Learning Meaning and Social Identity*, Cambridge: Cambridge University Press

Whitty, G. and Willmott, E. (1995) 'Competence-based teacher education: approaches and issues', in Kerry, T. and Shelton Mayes, A. (eds) (1995) *Issues in Mentoring*, London: Routledge, pp. 225–31

Wilkin, M. (1992) *Mentoring in Schools*, London: Kogan Page

Wilkin, M. (1999) *The Role of Higher Education in Initial Teacher Education*, Occasional Paper No. 12, London: Universities Council for the Education of Teachers

Young, M. (1998) *The Curriculum of the Future*, London: Falmer

Index

Note: page numbers in *italics* denote tables or figures.

Aberdeen University 83n1
accountability 3, 4, 34, 53, 78
Anderson, R. xi
apprenticeship model 34–5
assessment methods 6, 55–6, 60, 63, 67; self-assessment 47

Bachelor of Education (BEd) 29, 74, 76
Baistow, K. 17
Ball, S. 3, 19, 21, 54, 82
Beck, U. 3
Bertaux, D. 16
Bowe, R. 19
Cameron-Jones, M. 56
Campbell, A. 37
Chartered Teacher programme 26
child-centredness 50
Clark, K. 30
classroom management skills 44, 48, 49, 67
Collaborative Review proposals 34, 39
colleges of education 21–3
communication skills 45
comparative research methods 7–8, 9–10, 13, *14*, 15
competences 24, 39, 52, 53
compliance, culture of 50
Concise Oxford Dictionary 37
Conservative government 27, 81
context sensitivity 7, 9, 82–3
Convergence or Divergence study 3, 4, 55, 65–71, 73–4
Council for the Accreditation of Teacher Education 16, 28–9, 64
Crossley, M. 5, 7

Dale, R. 21
data collection 9, 10–15
Deloitte & Touche 27, 58
Department for Education and Skills 8, 16, 30, 33–4
devolution: divergence 4–5, 78; education xi, 26–8, 72, 80; globalisation 2
Director of Education 26
discourse of derision 81

discourse/practice 8, 13–14, 42, 46–52
Dundee University 75

Education, Governance and Social Integration and Exclusion in Education project 4
Education Act (1994) 64
education policy xi, 1–2, 20–1, 72, 79–81
Education Reform Act (1988) 2, 19
Educational Institute of Scotland 25–6, 35
employment-based routes 48–9, 51, 76, 77, 78
England: education system 6; educational policy 20–1; GTP 9–10; institutional stakeholders 8–9; ITT 28–34, 48–51, 76–7; key documents 38, 41–4; National Curriculum 1–2, 6, 42, 43; partnership 57, 63–72; policy community 20–1, 33–4; policy trajectory 34–5; recruitment levels 28, 31; retention 31
Eraut, M. 37
European Commission 3, 4, 45, 73
Eurydice report 3, 45

fieldwork practice *14*, 15
Furlong, J. 55, 56, 63, 65

Gauthier, P.-L. 4
Geertz, C. 13
General Teaching Council for England 8, 30, 33, 43, 58
General Teaching Council for Scotland xi, 8, 23, 25, 72, 74
globalisation 2–3, 4, 5, 6, 79–80
Goodson, I. 37–8
Grace, G. 18–19
Graduate Teacher Programme 9–10, 32, 33, 34–5, 50, 70–1
Green Paper on teachers 64, 79
Guidance Notes 38, 64
Guidelines for Initial Teacher Education Courses in Scotland 38, 40

Handbook of Guidance 38, 44
Her Majesty's Inspectorate (HMI) 31–3
Her Majesty's Inspectorate in Education (HMIE) 8, 16, 23, 24, 34, 74

Index

Higher Education Institutes 25, 35; Aspect Reviews 25; ITE 59–60; ITT 29, 30, 34, 64–5; partnership 61–2; staff trainers 49–50
Higher Education tutors 55–6, 67, 69–70
Higher Level Teaching Assistants 77
historical perspective 7, 8–10, 18–19
Holmes, H. xi
Humes, W. 9, 20

initial teacher education, Scotland (ITE) 3, 17n1, 21–8; HEI staff 59–60; homogeneity 74–5; ITT 45; school staff 60–1; stakeholders in partnership 75
initial teacher education and training (ITET) 1–2, 3, 17n1; convergence/divergence 4, 6–8; HEIs 35; historical perspective 8–10; home international study 5–6; Key Moments 10–11, *12,* 13; key policy documents 38; partnership 55–7; policy making 8–9; policy trajectories 34–5; Scotland and England compared x–xii, 1, 3–5, 18, 46, 52–4, 77–8; themes 78–9
initial teacher education and training providers 8, 33, 43–4, 46–7, 48–51
initial teacher training, England (ITT) 17n1, 28–34; diversification 16–17, 29–31, 76–7; Guidance Notes 64; HEIs 34, 64–5; ITE 45; quality issues 31–3; requirements for 42–3
inspection 51, 68; *see also* Her Majesty's Inspectorate; OfSTED
interviewing 9, 65–71
Ireland, Northern 5–6, 83

Jordanhill College of Education 22

knowledge-based society 37
Kogan, M. 20

Labour government 27, 64; *see also* New Labour
Labour/Liberal Democrat administration 27
learning: cross-cultural 45; situated 37
learning society 54
local contexts 5, 36
Local Education Authorities 8, 26, 60, 65, 69

Maastricht treaty 3, 73
McCrone settlement 27, 35, 58, 79
McPherson, A. 9, 20
managerialism 82
Marker, W. 10
marketisation 42, 44–5
Mentor Teacher Initiative 27, 57, 58, 71–2
modernisation of teaching 53, 54

Moray House Institute of Education 57
Munro, N. 9

National Curriculum 1–2, 6, 42, 43
national identity 5, 81
National Partnership Project 65
neoliberalism 54, 82
New Labour 21, 42, 81
New Right 30
Newly Qualified Teacher 38

Office for Standards in Education (OfSTED) 8, 16, 29; compliance culture 50–1; inspection 44, 64, 68, 76; quality of provision 33, 53; TTA 68
O'Hara, P. 56
Olssen, M. 82
Organization for Economic Development (OECD) 3, 4, 73
Oxford Internship Scheme 57
Ozga, J. 3, 16

Paisley, University of 26
parallel development 79–80
Parker, S. 36, 53–4
partnership 3, 4; collaborative 56–7, 65–6; complementary 56, 57, 62, 63–4, 66; duplication 55, 57; England 57, 63–72, 79; HEI-led 56, 57, 61–2, 65, 69–70; integrated 55, 63; ITET 55–7; school staff 61, 71, 72; Scotland 57–63, 71–2, 79; separatist 56, 63–4; stakeholders 59–63, 75; university staff 62, 63; Working Group on 57–8
Paterson, L. xi, 2
personal research development 51
placements 55–6, 60, 61–2
policy community 18, 19–21, 33–4
policy cycle 19
policy trajectory 18–19, 34–5
Post Graduate Certificate in Education (PGCE) 27–8, 29, 50, 57, 76
practice: communities of 37; comparative research 13, *14,* 15; context sensitivity 9; fieldwork *14,* 15; policy 18–19; reflective 37, 47, 51–2, 56–7; theory 62–3
probationary teachers 60–1
professional development continuum 33, 45, 47, 50
Professional Graduate Diploma in Education (PGDE) 74
professional knowledge xii, 36, 37–8, 46, 52, 56–7
professionalisation 3–4, 34, 35, 78–9
professionalism 38, 43, 46, 61, 77, 81
pupil achievement comparisons 80

Qualified Teacher Status 31–3, 42, 43, 45, 76
Qualifying to Teach (TTA) 38, 41–5
Quality Assurance Agency (QAA) 39, 40–1
Quality Assurance frameworks 23, 34

Raab, C. D. 9, 20
Raffe, D. xi, 2, 5–6, 7
rationalisation 3, 4, 34, 78
recruitment levels 10, 27–8, 31, 81–2
registration for teaching 9–10, 27, 45
research methods 16, 46
retention 31
Ryle, G. 13

Sachs, J. 57
Schon, D. 37
school staff 35; ITE 60–1; ITT 29; partnership 61, 66, 69, 71, 72; professionalism 61; student teachers 66–7; *see also* professionalism
School-Centred Initial Teacher Training (SCITT) 29, 32–5, 48, 51, 64, 70–1
schools, self-government 6
schools' role 3, 29, 52, 67
Scotland: devolution xi, 2, 4–5, 26–8, 72, 78, 80; education system 5; ITE 21–8, 74–5; ITET providers 33, 46–7; key documents 38, 39–41; partnership 57–63, 71–2; policy community 19–21; policy trajectory 34–5; professional development continuum 45; recruitment levels 10, 27–8, 81–2; stakeholders in partnership 8–9, 59–63; *see also* professional knowledge
Scottish education myths 52, 83n4
Scottish Executive 5, 73; 'Building the best small country in the world' 81; Education Department 8, 23–5, 75; education policy 16, 35; established 5, 73; ITE 27 8; second stage review 82; *see also* McCrone settlement
Scottish Funding Council 24
Scottish Higher Education Funding 34
Scottish Office xi, 5, 22, 24, 39; Education and Industry Department 39, 40; Education Department 57, 75
Scottish Parliament 5, 73
Scottish Teacher Education Committee 33

Scottish Tertiary Education Advisory Council 21
secondary subject entry 44
Secretaries of State for Scotland xi, 5
social justice approach 21, 52
socio-historical approach 7
stakeholders in partnership 59–63, 75
Standard for Full Registration 27, 45
The Standard for Initial Teacher Education in Scotland 23, 34, 38, 40–1, 46–7
Standard for School Leadership (Headship) 45
Standing Conference of Principals 33, 65
Stenhouse, L. 46
Stirling Unversity 74, 75
Strathclyde University 22
structure/agency 21
student teachers 61–2, 63, 66–7

teacher educators 46–52, 51–2
Teacher Training Agency 8, 33, 64, 76; compliance culture 50; HEI manager 30; interviews 16; National Partnership Project 65; OfSTED 68; *Qualifying to Teach* 41–4
teacher-–pupil ratios 28
teachers' role in society 42, 44, 60, 81, 82
Teaching Quality Assessment 34
technicism 52
theory/practice 62–3
Thornton, K. 9
Tomlinson, S. 10
trade union 25–6, 33, 35, 60
Training and Development Agency for Schools 76
tutor assessments 63

Union, Treaty of xi, 5, 73
Universities Council for the Education of Teachers 33, 65
universitisation 4, 6, 22–3, 34, 47, 54, 78
university staff 51, 62, 63

Wales 5, 83
Walford, G. 9
Watson, K. 5, 7
Wilkin, M. 65

Young, M. 54

DUNEDIN ACADEMIC PRESS

POLICY AND PRACTICE IN EDUCATION 15

Series Editors:
Jim O'Brien and Christine Forde

This volume draws on a major study of initial teacher education and training (ITET) undertaken at the University of Paisley. It examines the key characteristics of ITET, including systems of governance, institutional arrangements, quality assurance processes, curriculum and assessment and the significance of ITET within national systems of education.

Further themes are the professional context of ITET in Scotland and in England and the roles of key stakeholders such as the government, schools and local authorities. The significance of recent political, social and cultural identities and their influence on the development of ITET policy and practice are considered. Finally the book looks at the ways in which ITET in these two countries is diverging, perhaps under the pressure of post-devolution nationalism, or converging, under the pressures of globalisation.

As a considered analysis of complex research findings this volume in the *Policy and Practice in Education* series will interest all those concerned with teacher education in Scotland and England. It raises questions of globalisation in education policy which will appeal to education policy makers in other countries considering the enhancement and development of their own teacher education and training provision.

Ian Menter holds the chair of teacher education at the University of Glasgow; **Estelle Brisard** is a lecturer in education at the School of Education, University of Paisley where **Ian Smith** is Dean of the School of Education.

About this Series
Education is a matter of critical concern to politicians, teachers and other professionals, parents and members of the general public. **Policy and Practice in Education** supports and illuminates the public and professional discussion of education and does so from a Scottish perspective set in its international environment.

Each volume in the series focuses on a particular aspect of education, reflecting upon the present and contemplating the future. The contributing authors are all well established and bring to their writing an intimate knowledge of their field, as well as the capacity to offer a readable and authoritative analysis of policies and practice.

Titles in the series include
4. Community Education, Lifelong Learning and Social Inclusion (2nd Ed)
5. Special Educational Needs (2nd Ed)
7. Values in Education – We're all Citizens Now
12. The Social Agenda of the School
13. Inter-Agency Collaboration – providing for children
14. Assessment
16. Induction – Fostering Career Development at all Stages
17. Gender and Teaching – Where have All the Men gone?

ISBN: 1-903765-47-1

9 781903 765470

DUNEDIN ACADEMIC PRESS

The Law of Damages

BREACH —— Causal connection —— DAMAGE

Stewart Dunn